Exploring Oracle Primavera P6 v7.0

CADCIM Technologies
525 St. Andrews Drive
Schererville, IN 46375, USA
(www.cadcim.com)

Contributing Author
Sham Tickoo
Professor
Purdue University Calumet
Hammond, Indiana, USA

CADCIM
Technologies
Excellence in Technology

CADCIM Technologies

Exploring Oracle Primavera P6 v7.0
Sham Tickoo

CADCIM Technologies
525 St Andrews Drive
Schererville, Indiana 46375, USA
www.cadcim.com

ISBN 978-1-936646-97-5

NOTICE TO THE READER

Publisher does not warrant or guarantee any of the products described in the text or perform any independent analysis in connection with any of the product information contained in the text. Publisher does not assume, and expressly disclaims, any obligation to obtain and include information other than that provided to it by the manufacturer.

The reader is expressly warned to consider and adopt all safety precautions that might be indicated by the activities herein and to avoid all potential hazards. By following the instructions contained herein, the reader willingly assumes all risks in connection with such instructions.

The publisher makes no representation or warranties of any kind, including but not limited to, the warranties of fitness for particular purpose or merchantability, nor are any such representations implied with respect to the material set forth herein, and the publisher takes no responsibility with respect to such material. The publisher shall not be liable for any special, consequential, or exemplary damages resulting, in whole or part, from the reader's use of, or reliance upon this material.

www.cadcim.com

DEDICATION

*To teachers, who make it possible to disseminate knowledge
to enlighten the young and curious minds
of our future generations*

*To students, who are dedicated to learning new technologies
and making the world a better place to live in*

THANKS

To employees of CADCIM Technologies for their valuable help

Online Training Program Offered by CADCIM Technologies

CADCIM Technologies provides effective and affordable virtual online training on various software packages including Computer Aided Design, Manufacturing, and Engineering (CAD/CAM/CAE), computer programming languages, animation, architecture, and GIS. The training is delivered 'live' via Internet at any time, any place, and at any pace to individuals as well as the students of colleges, universities, and CAD/CAM/CAE training centers. The main features of this program are:

Training for Students and Companies in a Classroom Setting

Highly experienced instructors and qualified Engineers at CADCIM Technologies conduct the classes under the guidance of Prof. Sham Tickoo of Purdue University Calumet, USA. This team has authored several textbooks that are rated "one of the best" in their categories and are used in various colleges, universities, and training centers in North America, Europe, and in other parts of the world.

Training for Individuals

CADCIM Technologies with its cost effective and time saving initiative strives to deliver the training in the comfort of your home or work place, thereby relieving you from the hassles of traveling to training centers.

Training Offered on Software Packages

CADCIM Technologies provide basic and advanced training on the following software packages:

***CAD/CAM/CAE**: CATIA, Pro/ENGINEER Wildfire, SOLIDWORKS, Autodesk Inventor, Solid Edge, NX, AutoCAD, AutoCAD LT, Customizing AutoCAD, AutoCAD Electrical, EdgeCAM, and ANSYS*

***Architecture and GIS**: Autodesk Revit Architecture, AutoCAD Civil 3D, Autodesk Revit Structure, AutoCAD Map 3D, Autodesk Navisworks, and Autodesk Revit MEP, Bentley STAAD.Pro, Oracle Primavera P6*

***Animation and Styling**: Autodesk 3ds Max, 3ds Max Design, Autodesk Maya, Autodesk Alias, Pixologic ZBrush, and CINEMA 4D*

***Computer Programming**: C++, VB.NET, Oracle, AJAX, and Java*

For more information, please visit the following link:
http://www.cadcim.com

Note
If you are a faculty member, you can register by clicking on the following link to access the teaching resources: ***http://www.cadcim.com/Registration.aspx***. The student resources are available at ***http://www.cadcim.com***. We also provide **Live Virtual Online Training** on various software packages. For more information, write us at *sales@cadcim.com*.

Table of Contents

Chapter 1: Getting Started with Primavera P6

Chapter 2: Creating Projects

Chapter 3: Defining Calendars and Work Breakdown Structure

Chapter 6: Risks and Issues, and Settings Baselines

Chapter 7: Project Expenses and Tracking Progress of Project

Chapter 8: Printing Layouts and Reports

This page is intentionally left blank.

Preface

Oracle Primavera P6 v7.0

Oracle Primavera P6 v7.0 is a project management software developed by Oracle. This software is primarily used for project management and helps the users to manage projects of any size. Primavera P6 handles large scale projects which are cloud based, robust, and easy to use solution for globally planning, managing and executing project. Primavera gives project managers and schedulers opportunity to manage highly sophisticated and multifaceted projects.

The **Exploring Oracle Primavera P6 v7.0** textbook explains the concepts and principles of project management through practical examples, tutorials, and exercises. This enables the users to harness the power of managing project with Oracle Primavera P6 for their specific use. In this textbook, the author emphasizes on planning, managing and controlling project, assigning resources and roles to project, producing schedule and resources reports and graphics.This textbook is specially meant for professionals and students in engineering, project management, and allied fields in the building industry.

As you go through this textbook, you will work on tutorials and exercises that can be used to build a complete project. Each of these tutorials and exercises, though complete in themselves, will be a step toward accomplishing the larger projects.

The main features of this textbook are as follows:

- **Project-based Approach**
  The author has adopted a project-based approach and the learn-by-doing theme throughout the textbook. This approach guides the users through the process of creating the designs given in the tutorials.

- **Tips and Notes**
  Additional information related to various topics is provided to the users in the form of tips and notes.

- **Learning Objectives**
  The first page of every chapter summarizes the topics that are covered in that chapter.

- **Self-Evaluation Test, Review Questions, and Exercises**
  The chapter ends with Self-Evaluation Test so that the users can assess their knowledge of the chapter. The answers to the Self-Evaluation Test are given at the end of the chapter.

Also, the Review Questions and Exercises are given at the end of the chapters and they can be used by the instructors as test questions and exercises.

• **Heavily Illustrated Text**
   The text in this book is heavily illustrated with about 200 screen capture images.

Symbols Used in the Textbook

Note

The author has provided additional information to the users about the topic being discussed in the form of notes.

Tip

Special information and techniques are provided in the form of tips that help in increasing the efficiency of the users.

Formatting Conventions Used in the Textbook
Please refer to the following list for the formatting conventions used in this textbook.

• Names of tools, buttons, options, tabs, and menu are written in boldface.	Example: The **Add row** tool, the **Projects** button, the **General** tab, the **Filter By** option and the **File** menu.
• Names of dialog boxes, drop-downs, windows, drop-down lists, areas, edit boxes, check boxes, details table and radio buttons are written in boldface.	Example: The **User Preferences** dialog box, the **Layout** drop-down in the **Projects** window, the **Project Name** edit box in the **Projects Details** table, the **Active** check box in the **Resources Details** table, and so on.
• Values entered in edit boxes are written in boldface.	Example: Enter **Harbour Pointe** in the **Project ID** edit box.
• Names of the files saved are italicized.	Example: *c03_CONS_Home_tut01*

Naming Conventions Used in the Textbook

Tool
If you click on an item in a toolbar of any window and a command is invoked to create/edit an object or perform some action, then that item is termed as **tool**.

For example:
Print tool, **Print Preview** tool, **Columns** tool
Bars tool

Figure 1 Tools in the toolbar of the **Projects** window

Button
The item in a dialog box that has a 3D shape like a button is termed as **button**. For example, **OK** button, **Cancel** button, **Apply** button, **Select** button and so on, refer to Figure 2.

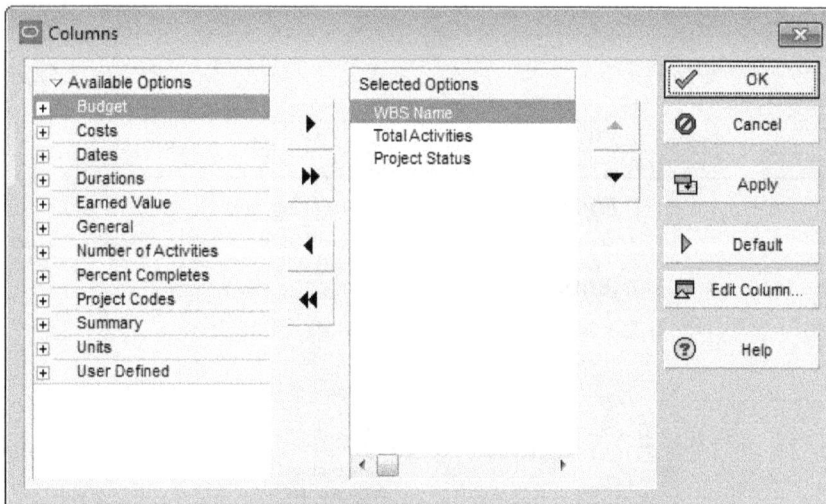

Figure 2 Choosing the **OK** button from the **Columns** dialog box

Dialog Box
In this textbook, different terms are used for referring to the components of a dialog box. Refer to Figure 3 for the terminology used.

Drop-down
A drop-down is the one in which a set of common tools and options are grouped together for creating an object or performing some action. You can identify a drop-down with a down arrow on it. These drop-downs are given a name based on the tools grouped in them. For example, **Layout** drop-down, **Display** drop-down, and so on; refer to Figure 4.

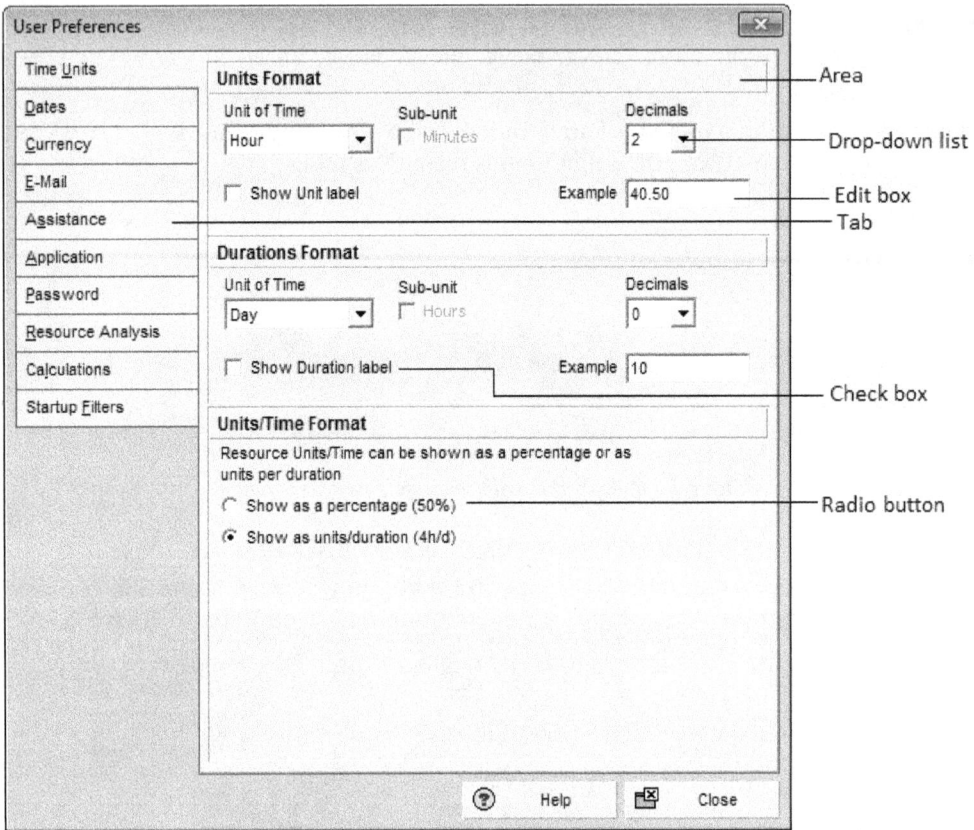

Figure 3 *Components of a dialog box*

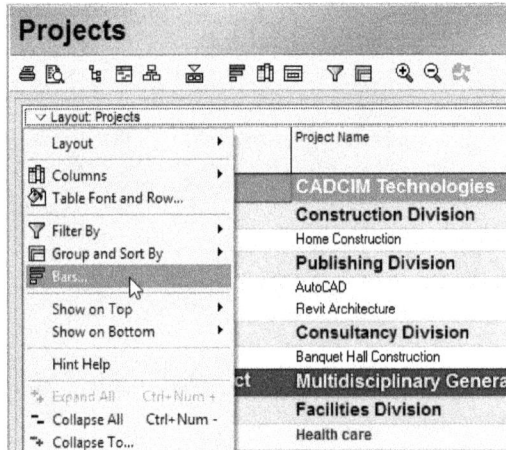

Figure 4 *Choosing an option from the drop-down*

Drop-down List

A drop-down list is the one in which a set of options are grouped together. You can set various parameters using these options. You can identify a drop-down list with a down arrow on it. For example, **Decimals** drop-down list, **Unit of Time** drop-down list, and so on; refer to Figure 3.

Options

Options are the items that are available in shortcut menus, drop-down lists, dialog boxes, drop-down lists, and so on. For example, choose the **Bars** option from the shortcut menu displayed on right-clicking in the Gantt chart area; refer to Figure 5.

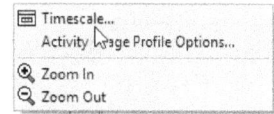

Figure 5   Choosing an option from the shortcut menu

Free Companion Website

It has been our constant endeavor to provide you the best textbooks and services at affordable price. In this endeavor, we have come out with a Free Companion website that will facilitate the process of teaching and learning of Oracle Primavera P6 v7.0. If you purchase this textbook, you will get access to the files on the Companion website.

The resources available for the faculty and students in this website are as follows:

Faculty Resources

• **Technical Support**
  You can get online technical support by contacting *techsupport@cadcim.com*.

• **Instructor Guide**
  Solutions to all review questions and exercises in the textbook are provided in this guide to help the faculty members test the skills of the students.

• **PowerPoint Presentations**
  The contents of the book are arranged in PowerPoint slides that can be used by the faculty for their lectures.

• **Primavera Files**
  The Primavera files (*.XER*) used in tutorials and exercises are available for free download.

Student Resources

• **Technical Support**
  You can get online technical support by contacting *techsupport@cadcim.com*.

• **Primavera Files**
  The Primavera files (*.XER*) used in tutorials and examples are available for free download.

• **Learning Resources**
  Additional learning resources at *http://pmpxperts.blogspot.com* and *http://youtube.com/cadcimtech*.

If you face any problem in accessing these files, please contact the publisher at *sales@cadcim.com* or the author at *stickoo@purduecal.edu* or *tickoo525@gmail.com*.

Stay Connected

You can now stay connected with us through Facebook and Twitter to get the latest information about our textbooks, videos, and teaching/learning resources. To stay informed of such updates, follow us on (*www.facebook.com/cadcim*) and Twitter (*@cadcimtech*). You can also subscribe to our YouTube channel (*www.youtube.com/cadcimtech*) to get the information about our latest video tutorials.

Chapter **1**

Getting Started with Primavera P6

Learning Objectives

After completing this chapter, you will be able to:
- *Understand the basics of project management*
- *Understand the need of project management*
- *Understand project planning*
- *Start Primavera P6*
- *Understand interface screen of Primavera P6*
- *Understand workspace*
- *Set the user and admin preferences*
- *Create a new project*
- *Open a project*
- *Export and Import a project*
- *Close project*

INTRODUCTION TO PRIMAVERA P6

Primavera P6 is a high end project management software which provides enterprise -wide solution. It helps in reducing costs, delivering the project on the projected time by streamlining, and coordinating with the management level changes. This software optimizes the overall resources so that the project runs efficiently no matter how big the size of the project is. It is designed to manage a large number of projects at a time to fulfill the project management requirements in an organization.

By using Primavera P6, one can simultaneously plan major project strategies and control the minute details to finish the project. The resources will be used effectively and productively.

In this chapter, you will be introduced to Primavera P6. Also, you will learn to navigate in this software and to customize and save screen layouts. With the completion of this chapter, you will be able to comfortably navigate through the Oracle Primavera P6 v7.0 and to customize the interface to suit your project needs.

BASIC FEATURES OF PRIMAVERA P6

Primavera P6 features a wide range of project management methodologies. These methodologies are briefly discussed below.

- **Centralized Project Repository**

Primavera P6 offers flexibility for determining who will access a project from all projects in a centralized database.

- **Enterprise Project Structure & Codes**

Enterprise Project Structure (EPS) helps to create a hierarchical structure for the projects based on the requirements of an organization.

- **Cross Project Analysis and Reporting**

You can easily and quickly create cross-project dependencies and can determine how projects affect one another. You can also determine whether the resources are under or over allocated across projects.

- **CPM Scheduling**

Primavera P6 provides Critical Path Method (CPM) scheduling using the activities and relationships between activities and calendars to maintain a project schedule. It identifies the critical activities that affect the completion of the project.

- **Float Path Analysis**

Float Path Analysis identifies all the critical paths within a project to avoid potential delays. It also helps in visualizing the activity's importance.

- **Cross Project Dependencies**

It helps in monitoring the overall critical path of activities and minimizes the risk of multiple parties working together.

*   **Resource Allocation**

Primavera P6 tracks labor, material, equipment, and expenses needed for the activities.

*   **ERP or Accounting Integration**

Primavera allows integration with ERP or accounting system so that the schedule and cost information can be shared.

*   **Resource Leveling**

It helps to ensure that sufficient resources are available to perform the activities in the project plan.

*   **Baseline Management**

Unlimited versions of the schedules, resources, and costs can be stored to compare how the project progresses with reference to the original plan.

*   **Project Reports**

In the Primavera P6, you can generate predefined and customized reports for a project.

CONCEPT AND NEED OF PROJECT MANAGEMENT

In simple terms, project management is the process of achieving set goals within the specified time, budget, and other resources. It helps to get maximum out of the available resources like manpower, money, materials, facilities, information, and many more. This process helps management or teams in an organization to plan, execute, and complete tasks on specific date or time and within a limited budget.

Following are three stages or phases of project management:
1. Planning
2. Controlling
3. Managing

PLANNING PROJECTS

Project planning is a part of project management related to the use of schedules, such as Gantt charts, that help to plan and manage projects in the project environment. While planning a project, first the scope of the project is defined by the duration of various tasks necessary to complete the work. These tasks are listed and grouped in Work Breakdown Structure. The duration of tasks is often estimated through an average of optimistic, normal, and pessimistic cases.

Second, the resources necessary for the given project can be estimated and the costs of activities can be calculated according to the resources assigned to those activities to get the total project cost. The project is then optimized to achieve the balance between the resource usage and project duration to comply with the objectives. This phase of project planning is followed by project schedule, and project management plan.

Purpose of Planning

The purpose of project planning is to achieve project objectives and to predict which activities and resources are critical for timely completion of the project. Therefore, planning helps in making strategies to attain the desired goals and thus ensuring that the project will be delivered on time and within the budget.

Planning helps in identifying the total scope of project and the resources to perform the work. It helps to optimize the resources according to the project needs. It also helps to identify the risks, plan to minimize them, and then set priorities of the project according to the risk. It also helps to provide a baseline plan based on which the progress of the project is measured.

Planning Cycle

When constructing a building, one of the first steps is to lay foundation. Similarly, before building-up projects using the Project Management module, one has to start from the foundation to be laid in the construction of Enterprise Project Structure. Figure 1-1 shows the flow chart of project planning.

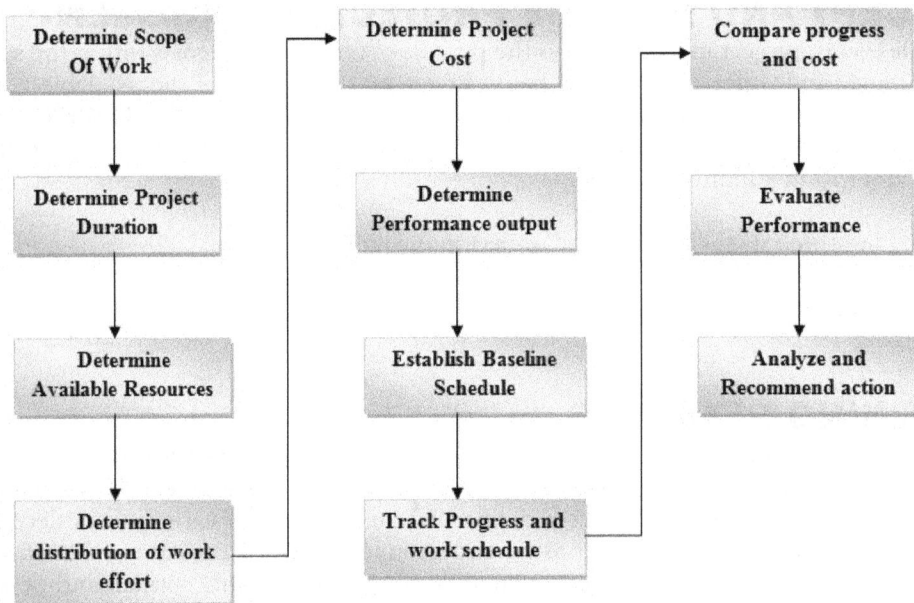

Figure 1-1 Flow chart of the project planning

The following steps are the detailed description of project planning using the Primavera P6 module.

• **Establish Goals**: The first step of the planning process is to identify and establish goals. The next step is to get the detailed information of goals including the reasons for the selection and the anticipated outcomes of the projects.

• **Established Goals to be achieved by Organization**: The Organizational Breakdown Structure (OBS) represents the hierarchical arrangement of the company management to fulfill the established needs. The OBS should be structured in such a way that each task in the Work Breakdown Structure (WBS) must be assigned to a person or committee. The OBS displays organizational relationships between the employees of an organization and according to that helps in assigning work resources in a project to the employees.

Each goal should have tasks or projects associated with its achievement. To achieve the goals, most useful and important structure for a project is the WBS. The WBS is continuation to the EPS for an individual projects in an organization. WBS can be identified as breaking down complex projects into simpler, manageable, and interrelated tasks. A good work breakdown structure encourages a systematic planning process that covers all the key elements in a project and simplifies the project into manageable units. A WBS is used as a road map for planning, monitoring, and managing all the elements of a project such as scope of work, cost and time estimation, resource allocation and scheduling, productivity and many a more.

• **Identifying Resources**: EPS is the hierarchical arrangements of the projects. It helps in managing multiple projects from higher to individual level and thereby facilitates the performance of specific tasks in an organization.

• **Prioritize Goals and Tasks**: Prioritizing goals and tasks process is to arrange goals or tasks in terms of their importance. The most important tasks will be theoretically and practically approached and completed first.

• **Goals Assigned to Resources**: Each goal has some financial and human resource requirements associated with it. To achieve the goals, certain resources are required to be assigned. In Primavera P6, Resources Breakdown Structure (RBS) is an hierarchical structure of resources which are needed to accomplish the objectives of project. It is logical and useful classification of resources that can be used to optimize resource utilization.

• **Create Assignments and Baselines**: As the projects are prioritized, it is required to establish baselines for completing associated tasks and assign resources to complete them. This portion of the management planning process should consider the abilities of the staff members and the time necessary to complete the assignments.

• **Tracking the Project's Progress**: A management planning process includes a strategy for evaluating the progress toward goal completion throughout an established time period. To evaluate the progress of a project, you need to keep a track of it.

CONTROLLING PROJECTS

Project control processes are performed to observe the project execution so that problems can be timely identified and corrective action can be taken, when necessary, to control the execution of the project. Project control includes measuring the ongoing project activities, monitoring the project variables such as cost, effort, scope and identifying the corrective actions to address risks and issues.

MANAGING PROJECTS

Project management is done by the Project Manager responsible for accomplishing the stated project objectives. Project management includes development of the project plan, managing project stakeholders, managing communication, managing project team, managing project risk, managing project schedule, managing project budget, managing project conflicts, and managing project delivery. Project Management depends upon the industry type, size, maturity, and culture of the company.

STARTING PRIMAVERA P6

Planning is an integral part in managing a project. Primavera P6 is a planning software that makes this activity an easier task. To start with Primavera P6, choose **Start > All Programs > Oracle - Primavera P6 > Project Management** (for Windows 7); the Primavera P6 v7.0 will open with the **Login to Primavera P6** dialog box, as shown in Figure 1-2. By default the login name is displayed in the **Login Name** edit. To login to Primavera P6, enter password in the **Password** edit box and then choose the **OK** button. By default, the login name and password to enter into primavera is **admin**.

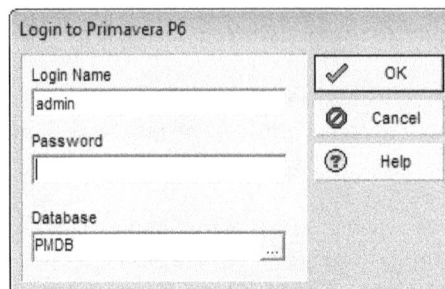

*Figure 1-2 The **Login to Primavera P6** dialog box*

A PMDB database is already created for primavera to work in the specified database. Create your own database by using the Browse button from the **Database** edit box. When you choose this button, the **Edit Database Connections** dialog box will be displayed, as shown in Figure 1-3.

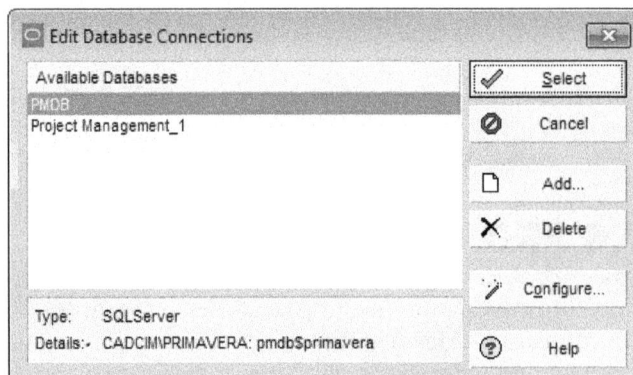

*Figure 1-3 The **Edit Database Connections** dialog box*

In this dialog box, select the required database from the **Available Databases** list box or continue working on the standalone primavera with predefined database. You can add more database to the primavera. To do so, choose the **Add** button from the Command bar of the dialog box; the **Database Configuration** wizard with the **Select or Create Alias** page will be displayed. Enter the required values in the page and keep choosing the **Next** button until you create your own database. As you finish creating database, choose the **Finish** button in this wizard. After specifying the database, choose the **OK** button in the **Login to Primavera P6** dialog box; the Primavera P6 user interface with the **Home** window will be displayed, refer to Figure 1-4.

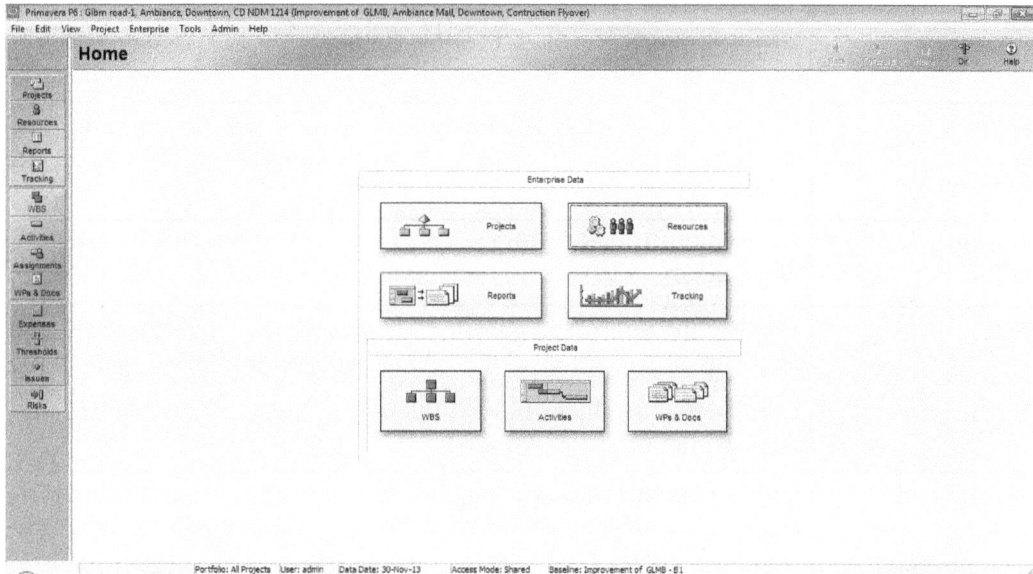

*Figure 1-4 The **Primavera P6** interface with the **Home** window*

USER INTERFACE SCREEN

The Primavera P6 interface with the **Home** window consists of Directory bar, menubar, Navigation Bar, and Status bar. You can hide and unhide these options using the toolbar. To hide any of the toolbars, choose **View > Toolbars**, refer to Figure 1-5 and then clear the check mark corresponding to the toolbar name that you want to hide in the **Home** screen.

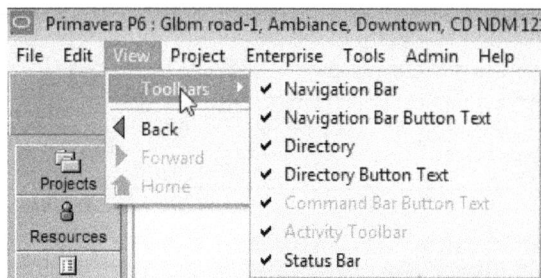

Figure 1-5 Options to hide or unhide a toolbar

Directory Bar

The buttons of the Directory bar help in displaying the windows quickly. The buttons in the Directory bar are described next.

Buttons	Description
Projects	Projects tab displays the **Project** window to review the Enterprise Project Structure (EPS) and allows to work on individual project.
Resources	Displays the **Resources** window. It helps to add and modify the organization resources.
Reports	Displays the **Reports** window which helps to create, run, edit, and delete project reports globally. You can also export and import the project reports.
Tracking	Displays the **Tracking** window which helps in monitoring the progress of projects.
WBS	Displays the **Work Breakdown Structure** window. This window helps in creating the work breakdown of the created projects.
Activities	Displays the **Activities** window in which you can create, edit, and modify the activities of an open project.
Assignments	Displays the **Resources Assignments** window to view all the resource assignments of opened projects. You can view, add, and delete the resources assigned to the activities.
WPs & Docs	Displays the **Work Products and Documents** window which helps to maintain work products and document records of the opened projects.
Expenses	Displays the **Project Expenses** window to work with all the expense items for the open project.
Thresholds	Displays the **Project Thresholds** window to calculate the estimated threshold for each activity in an open project.
Issues	Displays the **Project Issues** window which helps to add, edit, or delete project related issues.
Risks	Displays the **Project Risks** window which helps to calculate risks for the project.

Navigation Bar

The tools in the Navigation Bar, as shown in Figure 1-6 help to hide and unhide the Directory bar. It also provides access to the help file for the current window. The tools in the Navigation Bar are briefly described next.

Tool	Description
Back	Displays the last window opened by the user.
Forward	Displays the next window with a series of windows. Forward button is enabled only when you use the **Back** button to move to the next window.
Home	Displays the **Home** window.
Directory bar	Hide and unhide the Directory bar.
Help	Displays the **Help** window.

Figure 1-6 *The tools in the Navigation Bar*

Menubar

The menubar contains drop-down menus which are used to access functions like creating and opening project, setting user preferences, and so on. The menubar also helps to hide and unhide the toolbars. It also helps to access project activities, create WBS, EPS, OBS, and access the help file from the menubar. The menu of the menubar are briefly described next.

Menu	Description
File	It helps to open existing project or create new projects and also allows to print, import and export the project.
Edit	It helps to cut, copy, and paste the project and also helps in setting the user preferences.
View	It helps in the formatting of the projects.
Project	It includes all the elements of project such as activities, resource assignments, and WBS. It helps in creating WBS or activities and also enables users to enter into project risks and thresholds.
Enterprise	It helps in creating EPS, OBS, tracking projects, and arranging calendars.
Tools	It helps to generate reports and monitor thresholds.
Admin	It helps in the admin services.
Help	It allows to access the help content of primavera.

Command Bar

The Command bar is displayed at the right side when a project is opened. The Command bar displays the command to add a new project or activity, delete an existing project or activity and so on.

Status Bar

The Status bar is located at the bottom of the Project screen which describes the status of an opened project.

WORKSPACE

In the user interface, you can select any of the tabs to display the corresponding layout. The workspace of each window will display the menubar, Directory bar, Toolbar, Command bar and its detailed table. For example, if you choose the **Activities** tab from the Directory bar; the **Activities** window will be displayed, as shown in Figure 1-7.

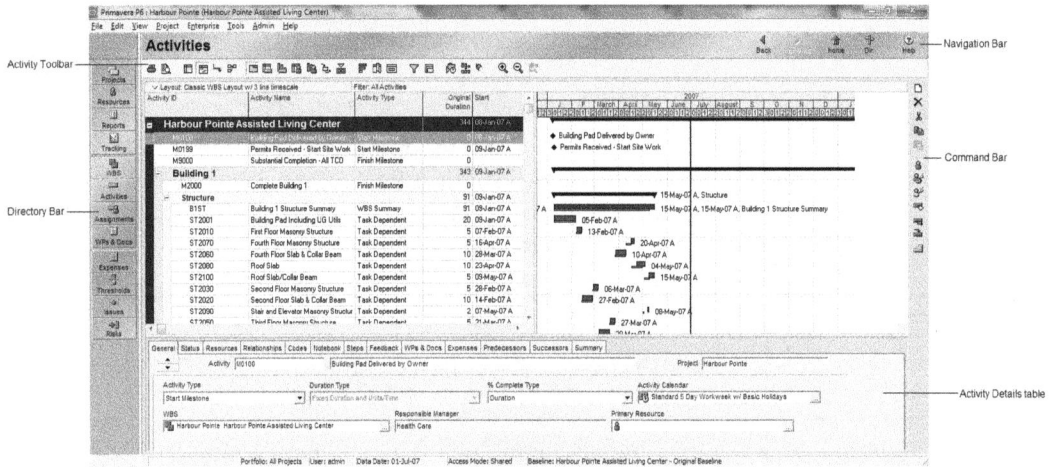

Figure 1-7 *The* **Activities** *window with the details of the window*

The **Details** table in the window displays the details of the opened window. You can enter general information, status, summary, comments, and so on related to the corresponding window in the table.

SETTING USER PREFERENCES

In Primavera, you can set user preferences as required. For example, indicate the format for displaying time units and dates, specify the currency to be used for viewing costs, and set startup display preferences. To set the user preferences, choose the **User Preferences** option from the **Edit** menu; the **User Preferences** dialog box will be displayed. This **User Preferences** dialog box is divided into ten tabs: **Time Units**, **Dates**, **Currency**, **E-mail**, **Assistance**, **Application**, **Password**, **Resource Analysis**, **Calculations**, and **Startup Filters**. The description of these tabs are given next.

Time Units Tab

The **Time Units** tab, as shown in Figure 1-8 allows you to define the time scale and to set the format that you want to use while displaying small and large-scale time units. It also allows to track layouts, set duration of activities, set resource prices, check availability, and display work efforts.

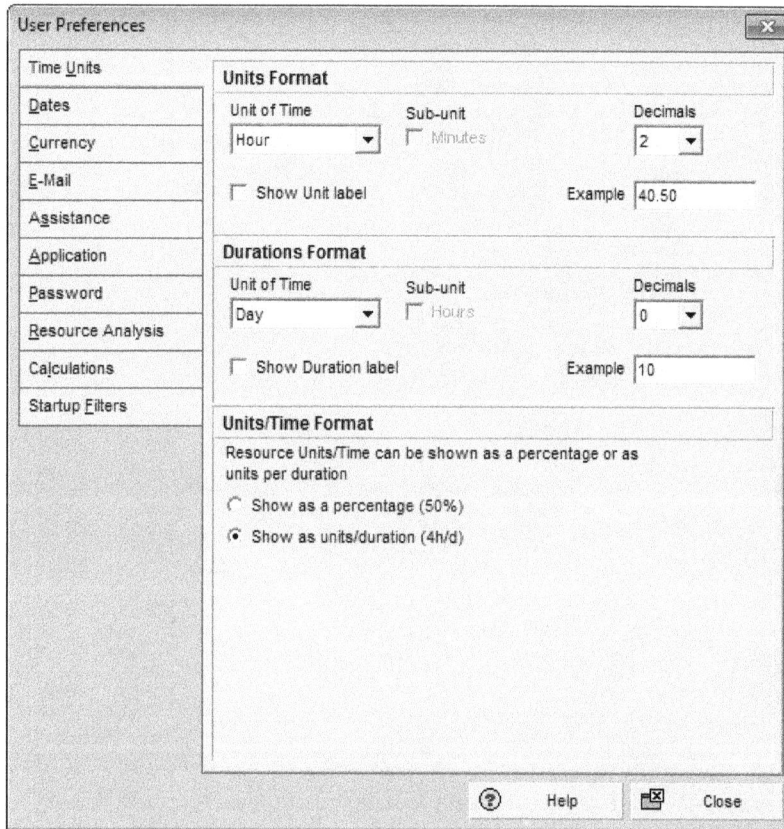

*Figure 1-8 The **User Preferences** dialog box with the **Time Units** tab displayed*

In the **Units Format** area, set the time unit in the **Unit of Time** drop-down list. If you want to include the next smallest time interval, select the **Show Unit label** check box. On doing so, the **Sub-unit** check box will be enabled. Select the **Sub-unit** check box to display the time with primary and secondary units. For example, if you select **Hour** in the **Unit of Time** drop-down list, the **Sub-unit** edit box will display **Minutes**. In the **Decimals** drop-down list, select the number of decimal places you want to include in the time unit value.

In the **Durations Format** area, select an option for the unit from the **Unit of Time** drop-down list. To enable the **Sub-unit** edit box, select the **Show Duration label** check box. In the **Decimals** drop-down list, select the number of decimal places to display the activity duration value.

In the **Units/Time Format** area, select the **Show as a percentage(50%)** radio button to display the resources units/time as percentage. To display the resources units/time as a unit duration value, select the **Show as units/duration** radio button.

Dates Tab

The **Dates** tab, as shown in Figure 1-9 allows to change the date and time format as desired.

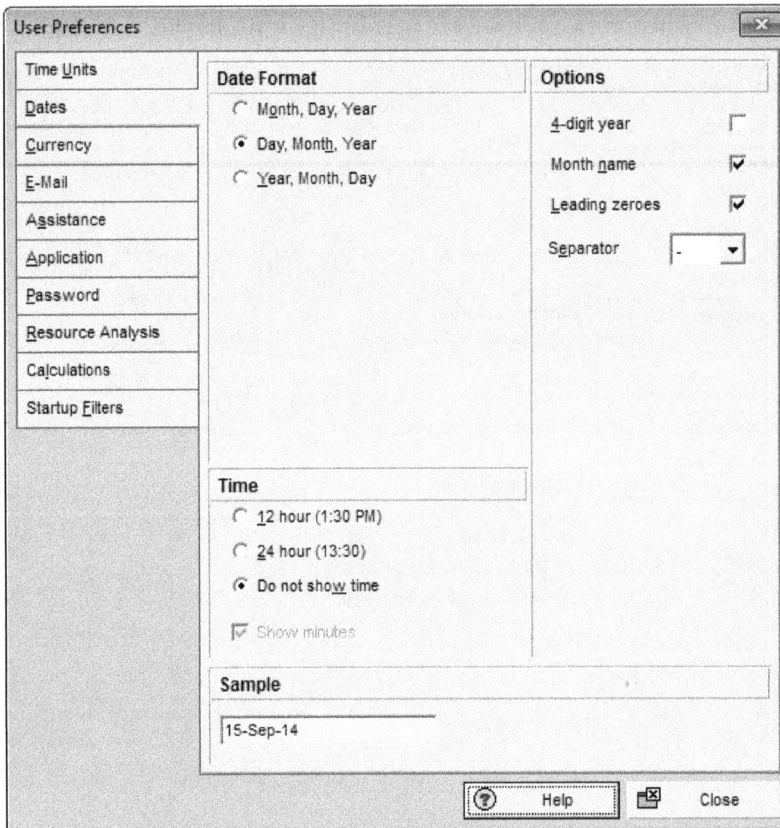

*Figure 1-9 The **User Preferences** dialog box with the **Dates** tab chosen*

In the **Data Format** area, select the radio button corresponding to the date format you want to use. Similarly, select the required radio button of the time format from the **Time** area. In the **Time** area, the **Show minutes** check box will be enabled only when you choose a radio button corresponding to time. Select the check boxes in the **Options** area to indicate how the selected date format should appear. In the **Separator** edit box of the **Options** area, select the character you want to use for distinguishing date, month and year.

Currency Tab

The **Currency** tab, as shown in Figure 1-10, allows you to select a currency for viewing monetary values. To edit the monetary values, choose the Browse button from the **Select a currency for viewing monetary values** edit box; the **Select Currency** dialog box will be displayed. In the **Select Currency** dialog box, select the desired currency and choose the **Select** button; the currency is assigned to the project and the **Select Currency** dialog box is closed and the **User Preferences** dialog box is displayed. The **Show currency symbol** and **Show decimal digits** check boxes are selected by default to display the currency symbol and decimal digit in the project.

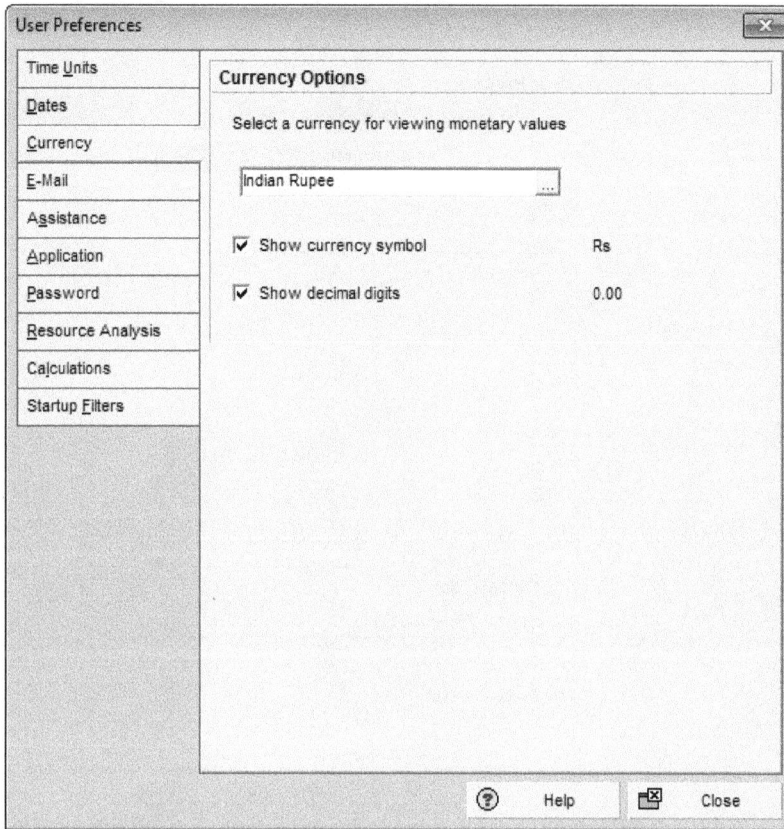

*Figure 1-10 The **User Preferences** dialog box with the **Currency** tab chosen*

E-mail Tab

The **E-mail** tab allows to enter the e-mail setting, as shown in Figure 1-11. It helps in specifying the mail protocol for an application e-mail. Select the type of protocol from the **E-mail Protocol** drop-down list. The module supports both the MAPI (Messaging Application Interface) and SMTP (Internet). By default, the **Internet** option is selected in the **E-mail Protocol** drop-down list and the **Mail Configuration** area is enabled. In the **Outgoing Mail Server (SMTP)** edit box, enter either the fully qualified domain name of the Internet mail server or its IP address. In the **User E-Mail Address** edit box, type the Internet mail address for the user from whom the mail will be sent. If you select the **MAPI** option from the **E-mail Protocol** drop-down list; the **Mail Server Login** area will be enabled. In the **Mail Server Login** area, enter the user profile name in the **Profile Name** edit box or retain the default set value. You can set the password for the user by choosing the **Password** button.

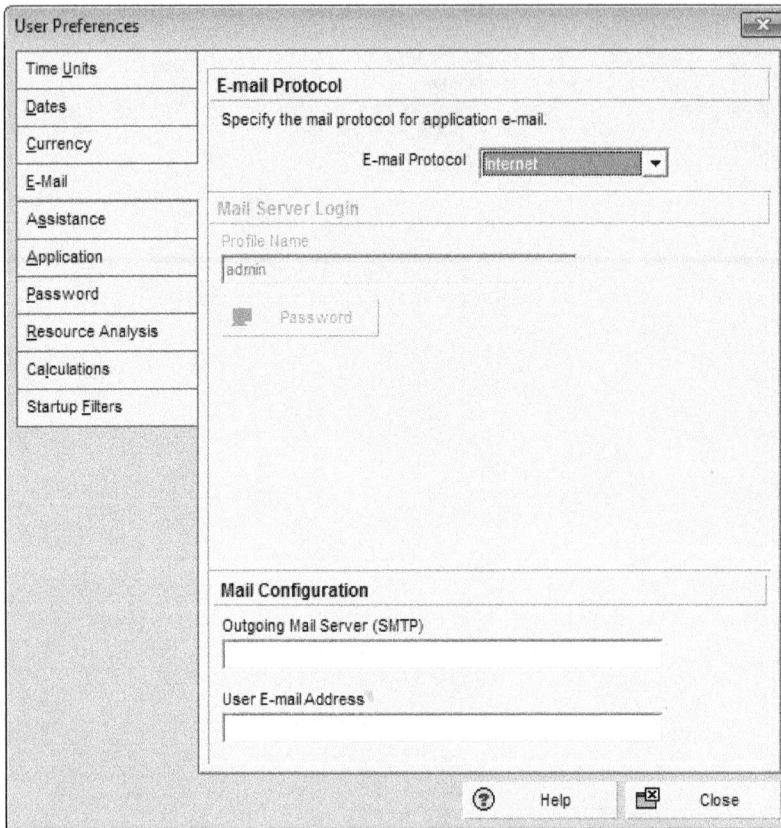

*Figure 1-11 The **User Preferences** dialog box with the **E-mail** tab chosen*

Assistance Tab

The **Assistance** tab, refer to Figure 1-12, allows to use wizards instead of standard dialog boxes. Wizards guide you through necessary steps to complete a function. As you become comfortable in adding resources and activities, you can easily add them and do not need the directions. Select the check boxes in the **Wizards** area to display the **New Resource Wizard** for adding a new resource, and the **New Activity Wizard** wizard for adding a new activity in their respective windows.

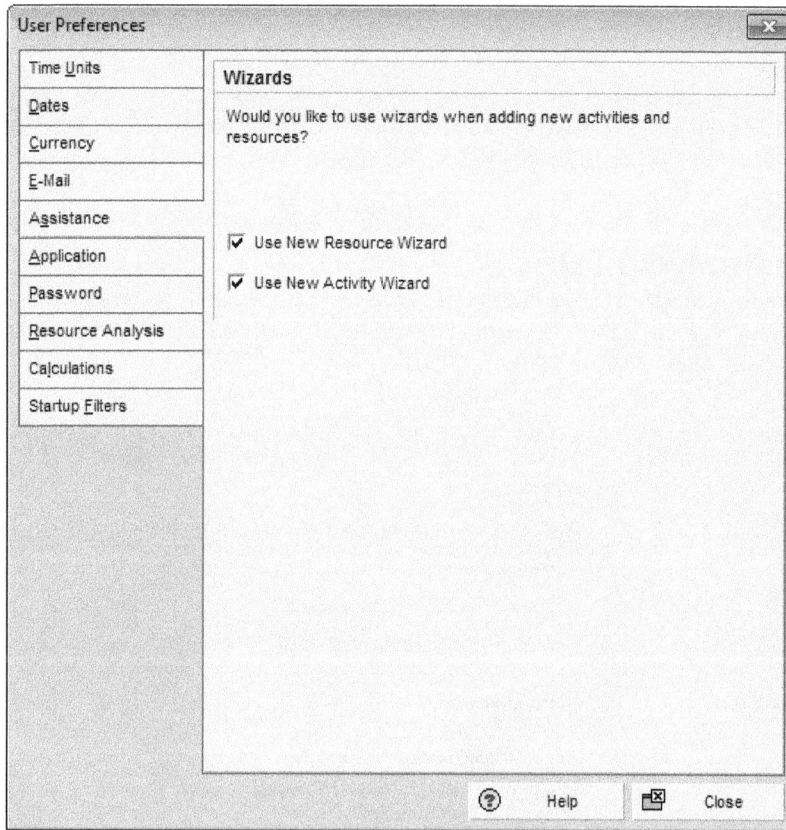

Figure 1-12 *The* **User Preferences** *dialog box with the* **Assistance** *tab chosen*

Application Tab

The **Application** tab allows you to customize startup window, and also enables you to trace the internal function to log file. You can set options for grouping and sorting. In the **Columns** area, you can set the financial periods to be viewed in columns. In the **Startup Window** area, select the options for the window that you wish to display each time you start the module, from the **Application Startup Window** drop-down list. Select the **Show the Issue Navigator dialog at startup** and **Show the Welcome dialog at startup** check boxes to display the **Issue Navigator** and **Welcome** dialog boxes at the startup, respectively. The **Issue Navigator** dialog box, contains outstanding issues that are generated based on your present thresholds. The **Welcome** dialog box enables you to choose to create a new project, open an existing project, open the last project opened in your previous session, and display global data only.

In the **Grouping and Sorting** area, you can enable or disable the **ID/Code** or **Name/Description** edit box by selecting or clearing the corresponding check boxes. In the **Columns** area, choose the Browse button to assign a range of financial periods to be viewed in columns.

Password Tab

The **Password** tab allows you to change the password of the current application. To change the password, choose the **Password** button; the **Change Password** dialog box will be displayed. In this dialog box, enter a new password and then confirm the password. Next, choose the **OK** button; the **Primavera P6** message box will be displayed informing you that password has been changed successfully.

Resource Analysis Tab

The **Resource Analysis** tab, refer to Figure 1-13 allows you to specify to be shown in the resource usage spreadsheet and the dates and time interval to be used to calculate time-distributed data for resource spreadsheets and profiles.

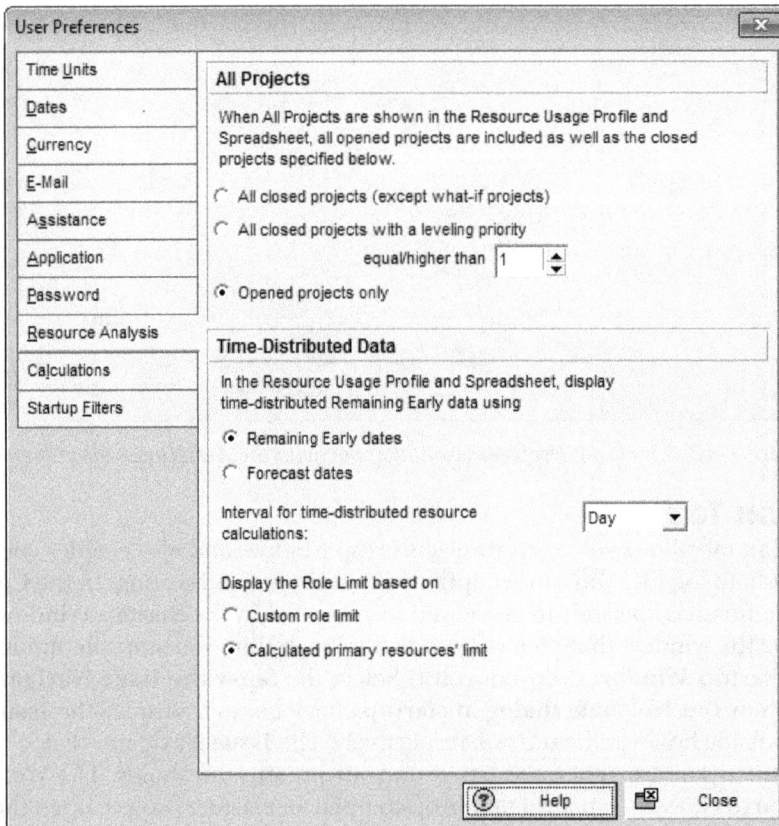

*Figure 1-13 The **User Preferences** dialog box with the **Resource Analysis** tab chosen*

In the **All Projects** area, you can select any one of the radio buttons to define the projects you want to be displayed in the resources usage profile and spreadsheet. In the **Time-Distributed Data** area, select a starting point for calculating remaining units and costs to be displayed in the Resource Usage Profiles, the Resource Usage Spreadsheet, tracking layouts, and Primavera Web application charts. To show the remaining values, select the **Remaining Early dates** radio button to calculate values based on remaining start/finish dates. To show remaining values, choose the **Forecast dates** radio button to calculate values based on forecasted start/finish dates.

Next, select an interval from the **Interval for time-distributed resource calculations** drop-down list at which the live resource and cost calculations are performed for Resource Usage Profiles and Resource Usage Spreadsheet.

Calculations Tab

The **Calculations** tab, refer to Figure 1-14, allows you to allocate cost and units on adding and deleting multiple project resources. It also enables you to specify how to calculate remaining values when new resource assignments are added or removed from activities. You can also choose the default behavior when replacing a resource/role on an existing activity assignment with a different resource/role.

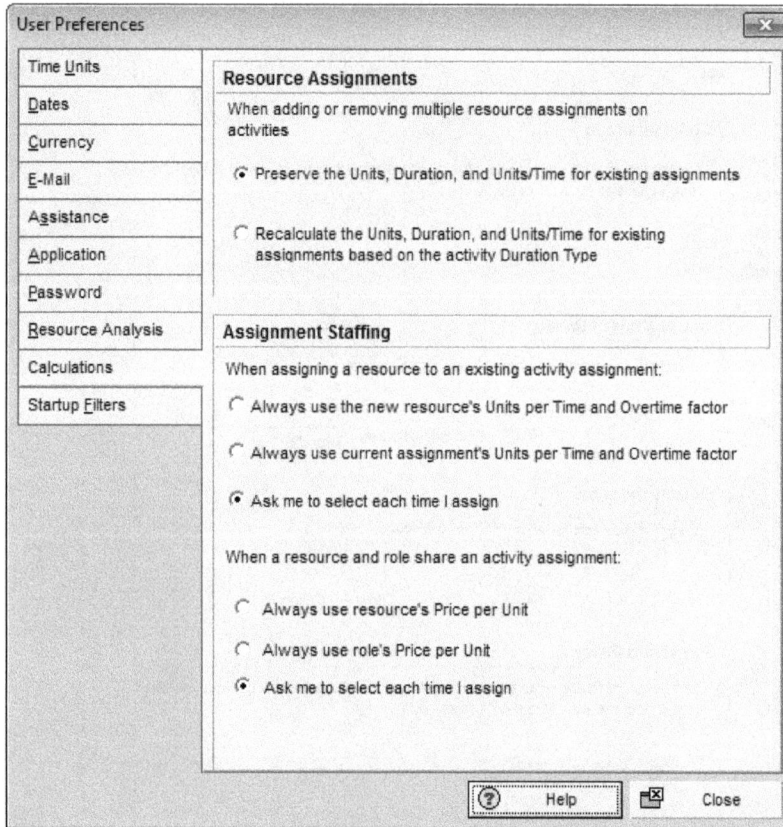

*Figure 1-14 The **User Preferences** dialog box with the **Calculations** tab chosen*

In the **Resource Assignments** area, select the **Preserve the Units, Duration, and Units/Time for existing assignments** radio button for units, duration and units/time to remain constant even if new resource assignments are added to existing projects. To recalculate the units, duration and units/time for existing assignments or for newly added assignments then select the **Recalculate the Units, Duration, and Units/Time for existing assignments based on the activity Duration Type** radio button. When replacing a resource on an existing activity assignment, you can choose to always use the units/time and overtime factor of the new resource. You can also specify the unit/time and overtime factor you want to use each time. When assigning a resource to an existing role, you can choose to always use the price/unit of the resource or role.

Startup Filters Tab

The **Startup Filters** tab contains the default filters to start the Primavera P6 application. You can choose this tab to view the data related to current projects or all the data in the database. These filters can reduce the time taken in opening projects.

SETTING ADMIN PREFERENCES

The admin preference settings established by the project controller are required to make specific changes in a project. To specify the admin settings, select the **Admin Preferences** option from the **Admin** menu; the **Admin Preferences** dialog box will be displayed, refer to Figure 1-15. This dialog box consists of nine tabs: **General**, **Timesheets**, **Data Limits**, **ID Lengths**, **Time Periods**, **Earned Value**, **Reports**, **Options**, and **Rate Types**.

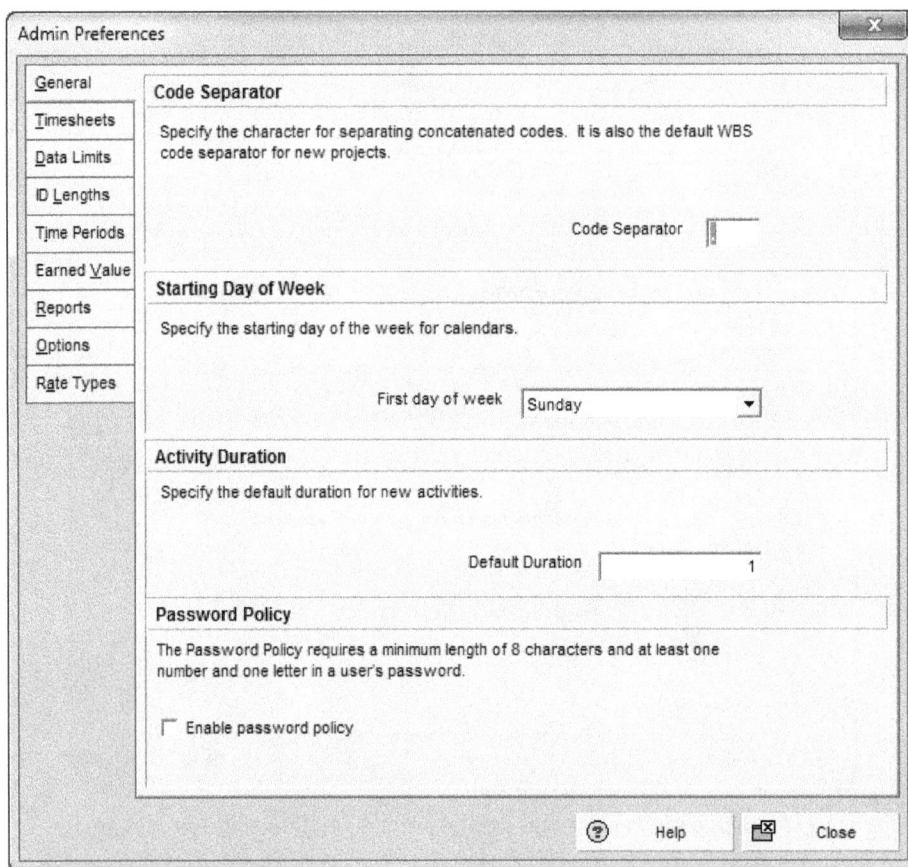

*Figure 1-15   The **Admin Preferences** dialog box with the **General** tab chosen*

General Tab

In the **General** tab, you can specify general options, such as the weekday from which the calendar week begins. You can change the character used to separate hierarchy levels in resource, project, activity codes, roles, cost accounts, and WBS elements. You can also change the default duration of activities.

Timesheets Tab

In the **Timesheets** tab, specify different options for the timesheet module. You can select the check box corresponding to the required options in the **General Settings** area. In the **Timesheet Approval Level** area, you can select any of the options to approve the timesheet filled during project completion. Primavera accepts the settings as made by project controller.

Data Limits Tab

The **Data Limits** tab enables you to specify the maximum number of levels for hierarchical structures. You can also specify the maximum number of baselines and activity codes that can be included in a project.

ID Lengths Tab

The **ID Lengths** tab enables you to set the number of characters to be displayed as ID or Code such as Project ID, WBS Code, Resource ID, Activity ID, Cost Account ID, and so on at each tree level.

Time Periods Tab

In the **Time Periods** tab, you can set the number of work hours such as Hours/Day, Hours/Week, Hours/Month, or Hours/Year. You can also specify abbreviations for displaying minutes, hours, days, weeks, months, and years.

Earned Value Tab

In the **Earned Value** tab, you can change the default earned value settings. You can also change the settings for computing the techniques of budget estimation.

Reports Tab

In the Reports tab, you can make settings for the display of reports. Using this tab, you can define the sets to display the reports. It can be displayed in first set, second set, and third set with the header, footer, and custom label for reports.

Options Tab

In the **Options** tab, you can specify the time interval on which cost and quantity summaries should be calculated for Resource Usage Spreadsheet and Activity Usage Spreadsheet. You can also make settings by selecting the **Allow use of Project Architect** check box to access whether users can access methodologies to add activities or create new projects using Project Architect.

Rate Types Tab

The **Rate Types** tab enables to specify the resources and roles rate types.

CREATING A NEW PROJECT

A project can be defined as a set of planned interrelated tasks to be executed over a fixed period of time and within certain cost and limitations. In Primavera P6, there are existing sample projects that can be used as a reference or you can create your own project.

WBS is defined as Work Breakdown Structure which is a hierarchical arrangement of deliverables produced during or by a project. WBS allows you to divide a project into meaningful and logical pieces for the purpose of planning and control.

The projects are created under EPS (Enterprise Project Structure). To create a new project, choose the **New** option from the **File** menu; the **Create a New Project** wizard will be displayed. In this dialog box, choose the Browse button corresponding to the **Select EPS** edit box and select the EPS under which you want to create the project.

OPENING A PROJECT

You can open a project by choosing the **Open** option from the **File** menu. When you choose this option, the **Open Project** dialog box will be displayed, refer to Figure 1-16. In this dialog box, expand the projects node to view the complete project. From the displayed list, select a project and choose the **Open** button to open the existing project.

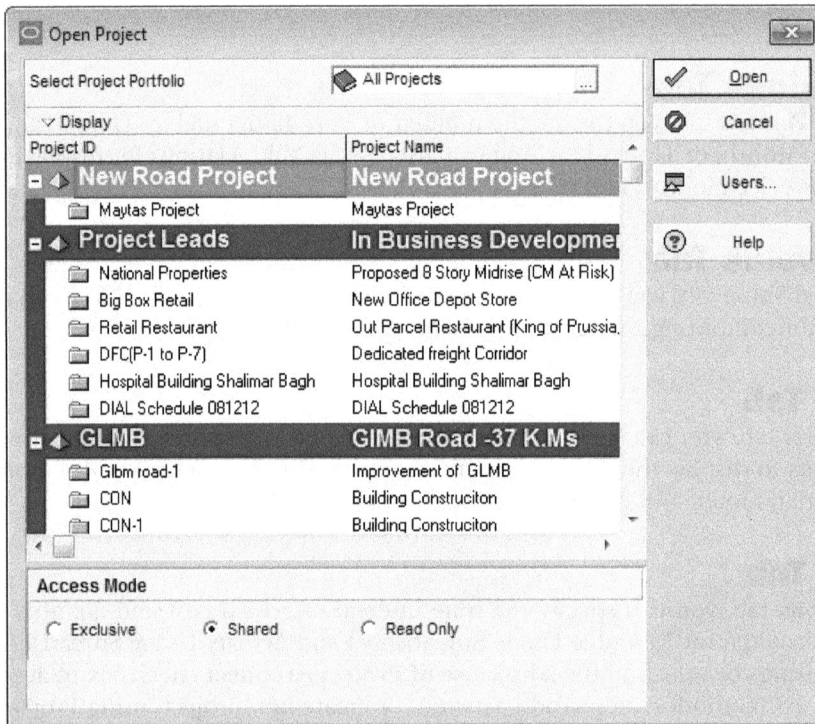

*Figure 1-16 The **Open Project** dialog box*

EXPORTING A PROJECT

You can export a project and save it for further use. To export a project, choose the **Export** option available in the **File** menu; the **Export** dialog box with the **Export Format** page will be displayed, as shown in Figure 1-17.

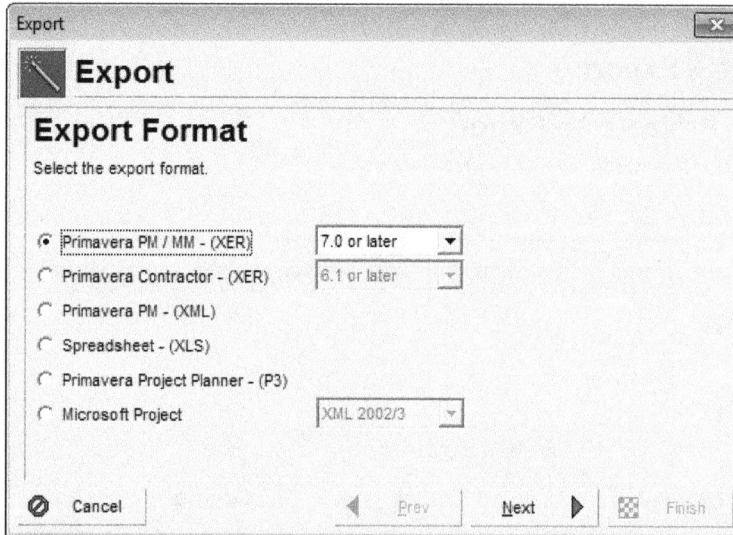

Figure 1-17 *The* ***Export*** *dialog box with the* ***Export Format*** *page*

In this dialog box, select the required radio button to set the format in which you want to export the file. Then, choose the **Next** button; the **Export Type** page will be displayed, as shown in Figure 1-18.

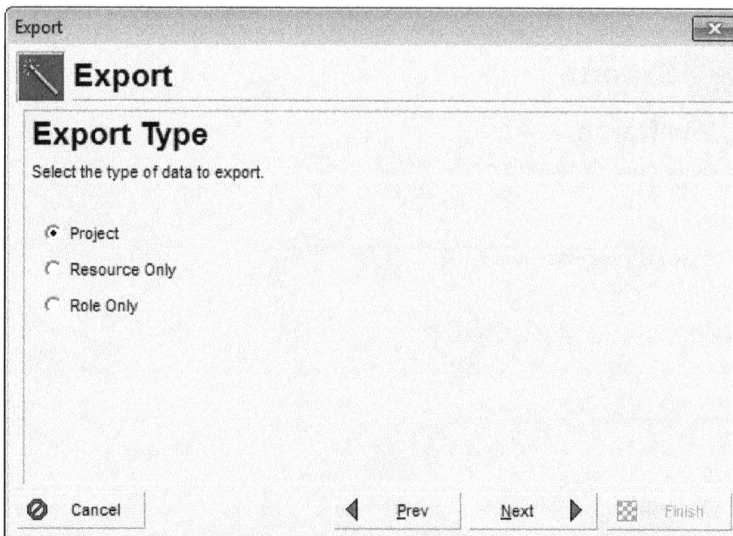

Figure 1-18 *The* ***Export*** *dialog box with the* ***Export Type*** *page*

In this page, select the type of data that you want to export by selecting the **Project**, **Resource Only**, or **Role Only** radio button. On selecting the **Project** radio button, the complete project will be exported. The **Resource Only** radio button allows you to export the resources of the open project. You can select the **Role Only** radio button to export the roles of the opened project. Now, choose the **Next** button; the **Projects To Export** page will be displayed, as shown in Figure 1-19.

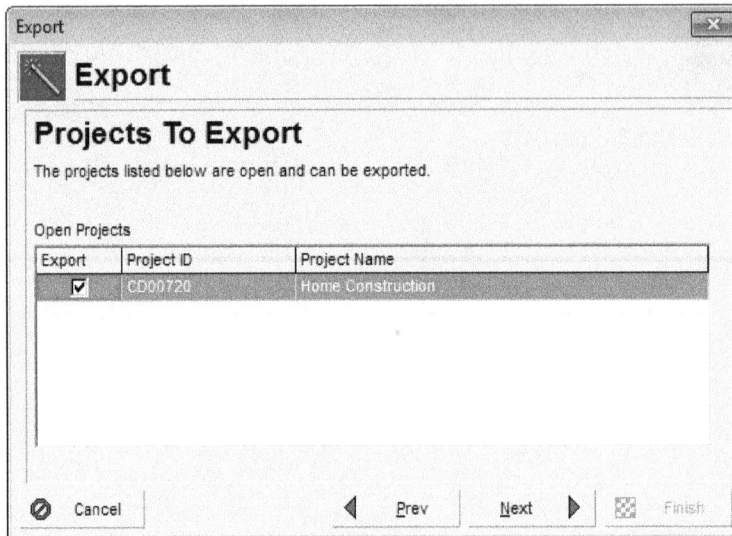

*Figure 1-19 The **Export** dialog box with the **Projects To Export** page*

In this page, you can view the opened projects that are to be exported as a file. Choose the **Next** button; the **File Name** page will be displayed, as shown in Figure 1-20. In this page, you can choose the Browse button corresponding to the **File Name** edit box.

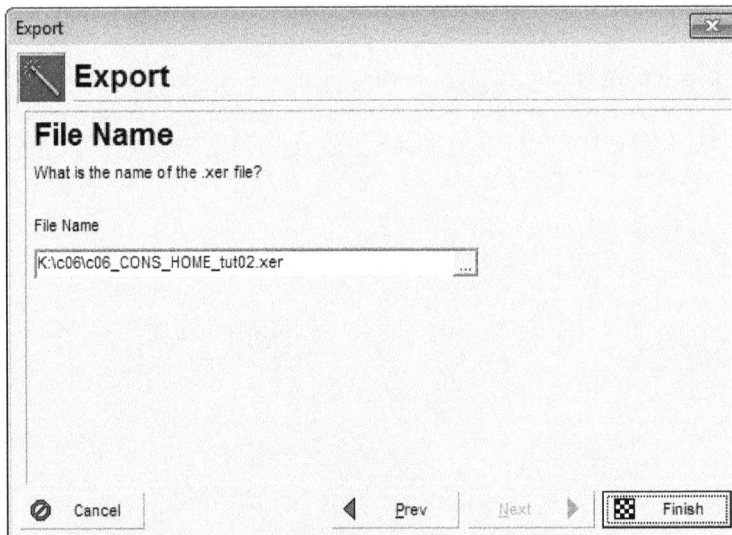

*Figure 1-20 The **Export** dialog box with the **File Name** page*

On doing so, the **Save File** dialog box will be displayed, as shown in Figure 1-21. In this dialog box, browse to the folder where you want to save the opened file and then choose the **OK** button; the **Save File** dialog box will be closed. Now choose the **Finish** button the export process will start and then the **Primavera P6** message box will be displayed, as shown in Figure 1-22, informing that the export was successful. Choose the **OK** button; the file will be exported and the export dialog box will be closed.

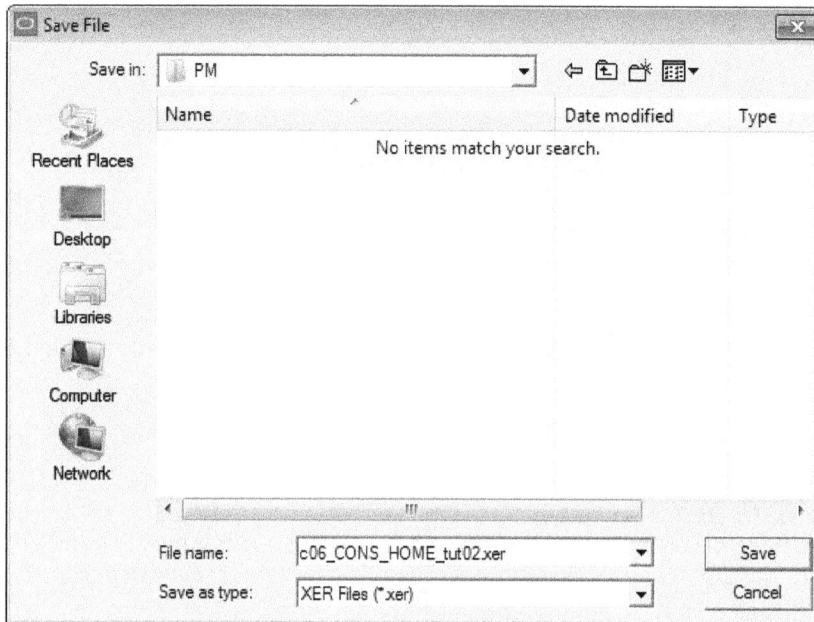

*Figure 1-21 The **Save File** dialog box*

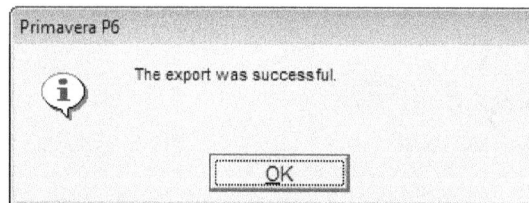

*Figure 1-22 The **Primavera P6** message box*

IMPORTING A PROJECT

To save time, you can import primavera files to the module. To import a project, Enterprise Project Structure (EPS) should be in the **Projects** window under which the project will be imported. Now, select the **Import** option from the **File** menu; the **Import** dialog box with the **Import Format** page will be displayed, as shown in Figure 1-23.

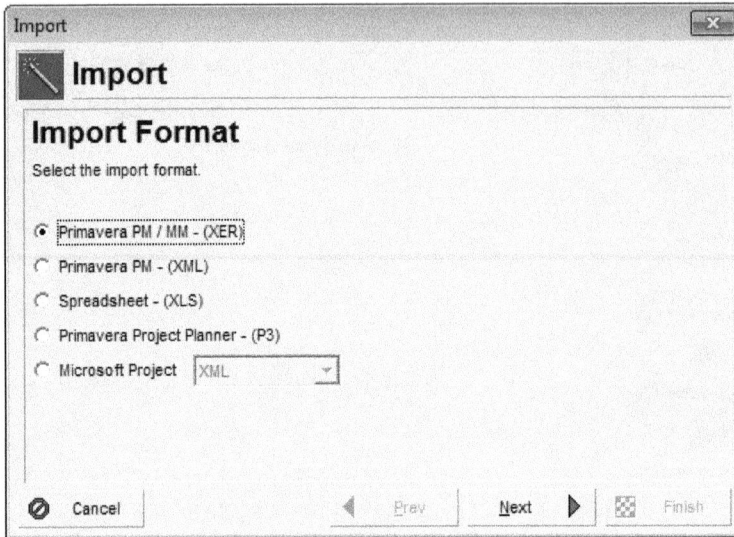

*Figure 1-23 The **Import** dialog box with the **Import Format** page*

In this dialog box, select the desired radio button for the file format that you want to import. Now, choose the **Next** button; the **Import Type** page will be displayed, as shown in Figure 1-24.

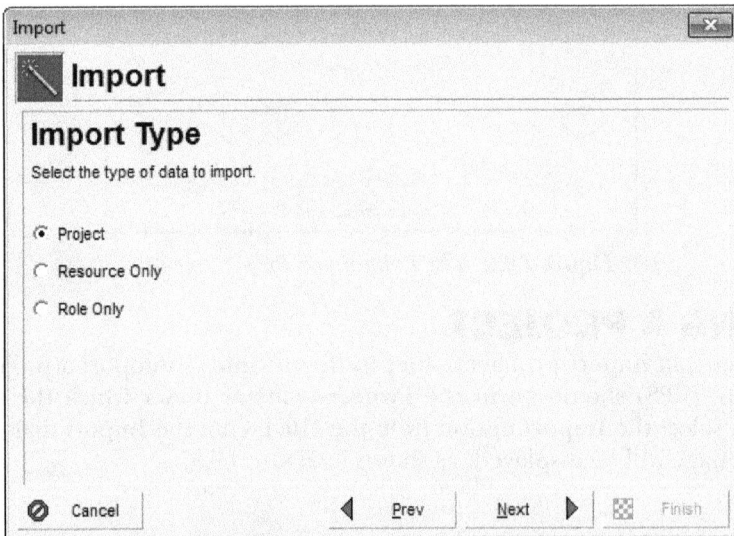

*Figure 1-24 The **Import** dialog box with the **Import Type** page*

In this page, select the radio button corresponding to the type of data you want to import in a project. The data types can be a project, resource or role only. After making the selection, choose the **Next** button; the **File Name** page will be displayed. In this page, browse to the *.xer* format file by choosing the Browse button corresponding to the **File Name** edit box; the **Select Import File** dialog box will be displayed. In this dialog box, browse to the location where the file is saved and then choose the **Open** button; the path of the file will be displayed in the **File Name** edit box. Choose the **Next** button; the **Import Project Options** page will be displayed, as shown in Figure 1-25. This page will allow you to find the project that is to be imported.

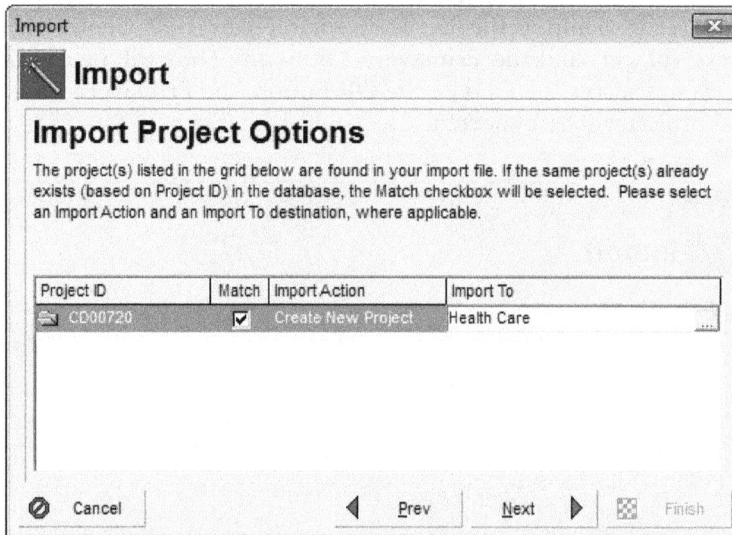

Figure 1-25   *The* **Import** *dialog box with the* **Import Project Options** *page*

Choose the Browse button under the **Import To** column; the **Select Project** dialog box will be displayed. In this dialog box, select the project that you want to import and then choose the **Select** button; the selected project will be assigned under the **Import To** column. Now, choose the **Next** button; the **Update Project Options** page will be displayed, as shown in Figure 1-26.

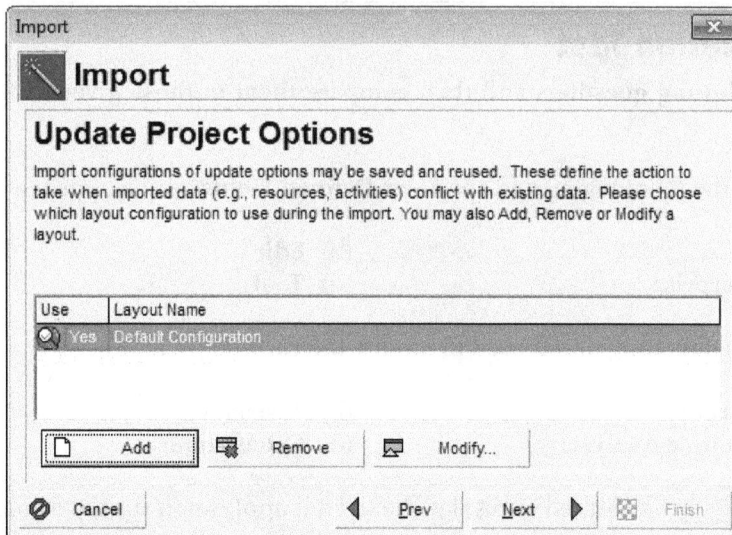

Figure 1-26   *The* **Import** *dialog box with the* **Update Project Options** *page*

In this page, you can add, modify, or remove the layout that is created in a project. To add a new layout, choose the **Add** button to make changes in the layout and name it accordingly. Once the modifications are done, choose the **Next** button; the **Finish** page will be displayed and will inform that you are ready to import the file, as shown in Figure 1-27. Choose the **Finish** button; the import process will start and the **Primavera P6** message box will be displayed informing you that the import was successful. Choose the **OK** button; the **Primavera P6** message box will be closed and the project will be imported.

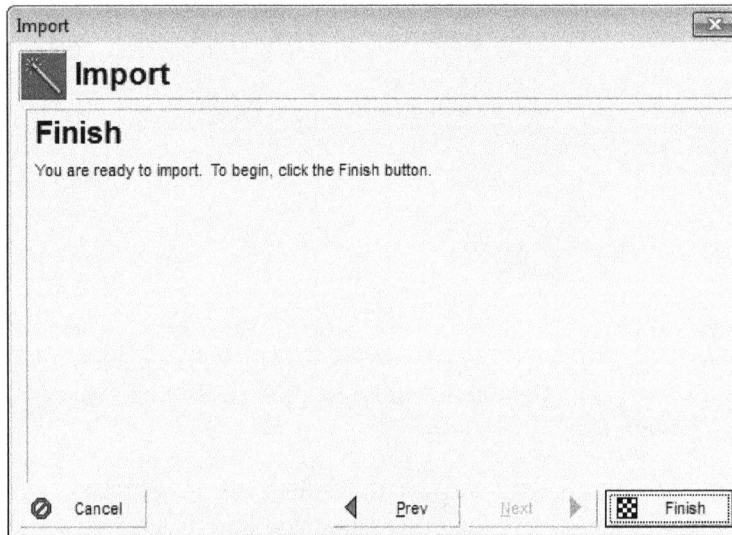

Figure 1-27 *The **Import** dialog box with the **Finish** page*

Self Evaluation Test

Answer the following questions and then compare them to those given at the end of this chapter:

1. In which of the following menus, the **User Defined Fields** option appears?

    (a) **File**   (b) **Edit**
    (c) **Enterprise**   (d) **Tool**

2. Which of the following tabs is used to modify the currency in a project?

    (a) **Currency**   (b) **Assistance**
    (c) **Resource Analysis**   (d) **Calculations**

3. The _____ tab is used to display the default application startup window.

4. The _____ window is used to calculate the threshold in a project.

5. The _____ tab is used to set the date format.

6. You can create your own database in Primavera P6 module. (T/F)

7. You cannot toggle on/off the display of toolbars. (T/F)

8. The **Time Units** tab enables you to change unit of time. (T/F)

9. In Primavera P6, you can access to either network such as MAPI or SMTP. (T/F)

10. You cannot control the display of wizards on adding new activities and resources. (T/F)

Review Questions

Answer the following questions:

1. Which of the following options allow you to import a file?

    (a) Export   (b) Page Setup
    (c) Send Project   (d) Import

2. Which of the following duration types is used in a project?

    (a) Fixed Unit   (b) Fixed Duration and Units/Time
    (c) Fixed Units/Time   (d) Fixed Duration & Units

3. Which of the following softwares is used to generate the project schedule?

    (a) Microsoft Excel   (b) Microsoft Project
    (c) Primavera   (d) All of the above

4. Which of the following is full form of MAPI ?

    (a) Messaging Application Internet   (b) Messaging Applicant Investigation
    (c) Mail Application Internet   (d) Messaging Application Interfcae

5. Which of the following tabs helps in displaying the issues in a project?

    (a) Risks   (b) Thresholds
    (c) Issues   (d) Expenses

6. Which of the following bars is displayed initial interface screen?

    (a) Navigation Bar   (b) Directory bar
    (c) Status bar   (d) All of the above

7. You can create a PMDB database in Primavera P6. (T/F)

8. You can open an existing project using the **Open** option from the **File** tab. (T/F)

9. You can create a new project using the **Open** option from the **File** menu? (T/F)

10. The **Issue Navigator** dialog box is used for outstanding issues that are generated based on the preset thresholds. (T/F)

Chapter 2

Creating Projects

Learning Objectives

After completing this chapter, you will be able to:

- *Understand the Enterprise Project Structure*
- *Create a project*
- *Copy and delete a project*
- *Understand the Project Details Window*
- *Understand the OBS*

INTRODUCTION

Planning is the first and most important function of management. It is required at every stage of management. Planning provides direction and facilitates decision making. It also reduces the risks of uncertainty, overlapping and wasteful activities. The business activities will not be successful in the absence of planning.

In this chapter, you will learn how to start planning and creating projects. Also, you will learn to structure and add projects to the hierarchy. The process of setting up the EPS and OBS and to use these methods for planning and managing project information is also explained in this chapter.

SETTING THE ENTERPRISE PROJECT STRUCTURE(EPS)

The EPS of an organization contains the hierarchical structure of all the projects in its database. The hierarchial structure is designed to show various levels required to represent the work on various projects in an organization. The EPS forms the initial grouping of the portfolios of the projects. It helps in managing multiple projects from the individuals to the highest levels of the organization. An organization has a large database that contains information about all the current projects and other related things. For efficient management of the projects within an organization the EPS should be structured in such a way that allows all the individuals to access the concerned project data.

The EPS can be further sub-divided into many nodes or levels as required. The EPS nodes and the procedure of adding the nodes are described next.

Understanding EPS Nodes

The EPS nodes represent the levels within the projects. The number of EPS nodes and their structure depends upon the scope of the projects and on the method of reviewing the data. The highest node in the EPS is the root node. The root node may have further divisions such as zones, area construction, supervising team and so on. Figure 2-1 shows an example of the EPS and the EPS nodes in which the **Innovative Construction** node is the EPS which includes a lower level EPS node called **Facilities**.

Figure 2-1 *The hierarchy showing nodes in a EPS*

The **Facilities** node is further divided into the **Health Care**, **Education,** and **Commercial** nodes. Each of these sub-nodes contains projects that are a part of the **Facilities** node. For example, the **Health Care** node contains the **Lincoln Hospital Project Rehab Center** and **Harbour Pointe Assisted Living Center** sub-nodes. You can add as many projects required to complete the desired work and which fall within the scope set by the operation executives and program managers or engineers in the organization. Multiple levels in the EPS helps you to manage projects separately while securing the data and summarizing it to the higher levels. For example, you can outline the data information up to each node in the EPS and also top-down budgeting can be checked from the top node to the lower node through their low-level projects for control.

Adding EPS Nodes

Menu Bar: Enterprise > Enterprise Project Structure

In Primavera P6, for assigning the EPS nodes to all the projects of an organization, choose the **Enterprise Project Structure** option from the **Enterprise** menu in the menubar; the **Enterprise Project Structure (EPS)** dialog box will be displayed, as shown in Figure 2-2. In this dialog box, select the node under which you want to add a new EPS node. Then, choose the **Add** button from the right pane of the dialog box; a new EPS subnode will be added under the selected node. You can also assign a unique ID and name to the new subnode under the **EPS ID** and **EPS Name** columns, respectively, refer to Figure 2-2.

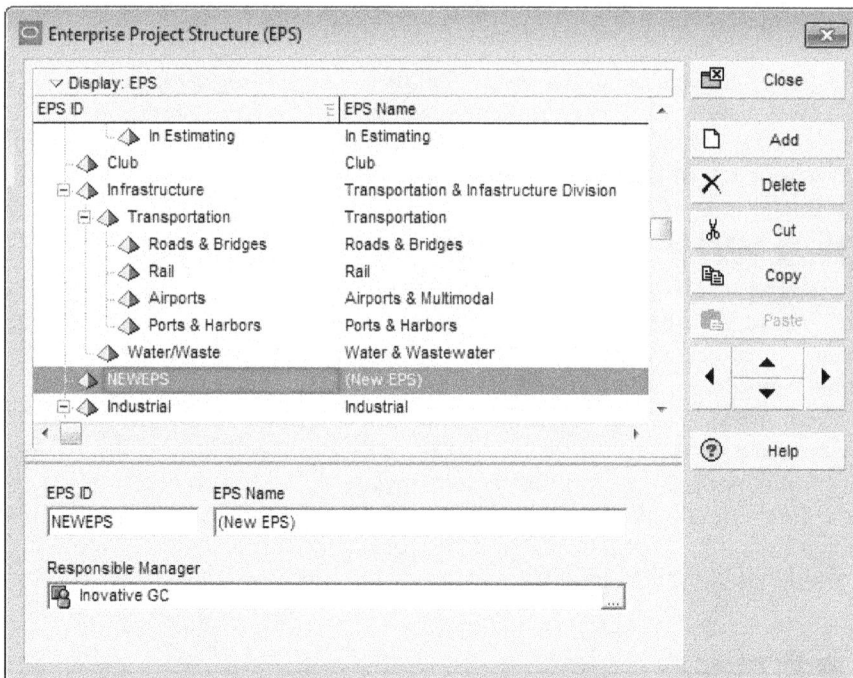

*Figure 2-2 The **Enterprise Project Structure (EPS)** dialog box*

In the **Responsible Manager** edit box, specify a responsible manager for the created node or choose the Browse button adjacent to it and select a different OBS element for the node. You can use the arrow keys available in the right pane of the **Enterprise Project Structure (EPS)** dialog box to indent/outdent a node to mark its placement in the EPS or to move a node up or down in the hierarchy.

You can categorize many levels in the EPS hierarchy by including more than one root node. To do so, add an EPS root node in the same way as you added an EPS node, but outdent the root to the left most position in the hierarchy. Choose the **Close** button to exit the dialog box; the **Projects** window will be displayed. In this window, the EPS will be displayed, as shown in Figure 2-3. The EPS nodes that contains subnodes and projects are identified by a pyramid symbol and further if there are more nodes or projects rolled up beneath the selected subnodes then the EPS will be symbolized by an addition symbol. Click on this symbol or double click on the node to display the additional nodes or levels in the hierarchy.

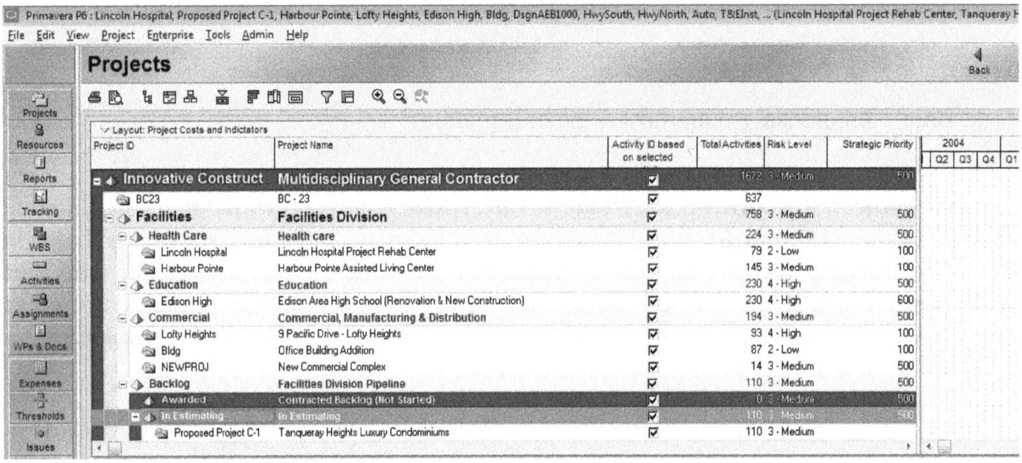

Figure 2-3 *The **Projects** window displaying the **EPS** containing the **EPS** nodes and sub-nodes*

You can open an EPS node to open all the projects in it or you can open the projects individually. In the next section, you will learn to work with the project and also to add a project.

WORKING WITH PROJECTS

A project is a temporary group activity which is designed to perform different tasks and constitutes a plan for creating a product or service. Being temporary, all the projects have a definitive start and finish dates. A project can involve a single person, a single organizational unit, or multiple organizational units.

An organization may have several projects to work on and each project has its own resource assignments. Similarly, each project has its own specific calendars, reports, and activity codes. Working as a program manager in an organization, an individual has to manage one or more higher level projects in the organization. Therefore, it would become convenient, if all such projects are added under an EPS node.

Adding a Project

To add a new project, open the **Projects** window by choosing the **Projects** tool from the **Enterprise** menu in the menubar. Alternatively, you can choose the **Projects** option from the **Home** window. In the **Projects** window, select the EPS node to which you want to add the project and then choose the **Add** button from the Command bar; the **Create a New Project** wizard with the **Select EPS** page will be displayed, as shown in Figure 2-4.

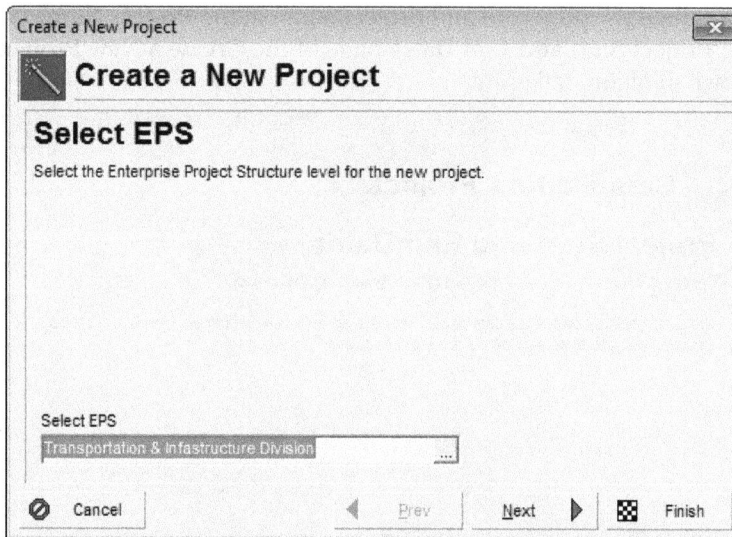

Figure 2-4 The Select EPS page of the Create a New Project wizard

In the **Select EPS** page of the **Create a New Project** wizard, the selected EPS name will be displayed in the **Select EPS** edit box. You can change the existing EPS to the required EPS by choosing the Browse button. On doing so, the **Select EPS to add into** dialog box will be displayed. In this dialog box, select the EPS to be assigned and then choose the **Select** button from the right pane of the dialog box; the **Select EPS to add into** dialog box will be closed and the selected EPS will be assigned to the project.

Now, choose the **Next** button; the **Project Name** page of the **Create a New Project** wizard will be displayed, as shown in Figure 2-5. In this page, you can specify the id and name of the project in the **Project ID** and **Project Name** edit boxes, respectively. Note that the project id is a short and unique identifier of the project.

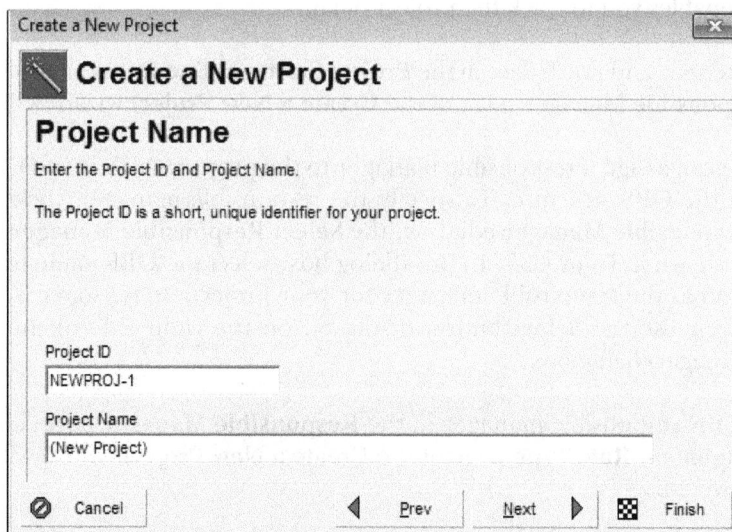

Figure 2-5 The Project Name page of the Create a New Project wizard

Once you have specified the project id and project name in the **Project Name** page, choose the **Next** button; the **Project Start and End Dates** page of the **Create a New Project** wizard will be displayed, as shown in Figure 2-6.

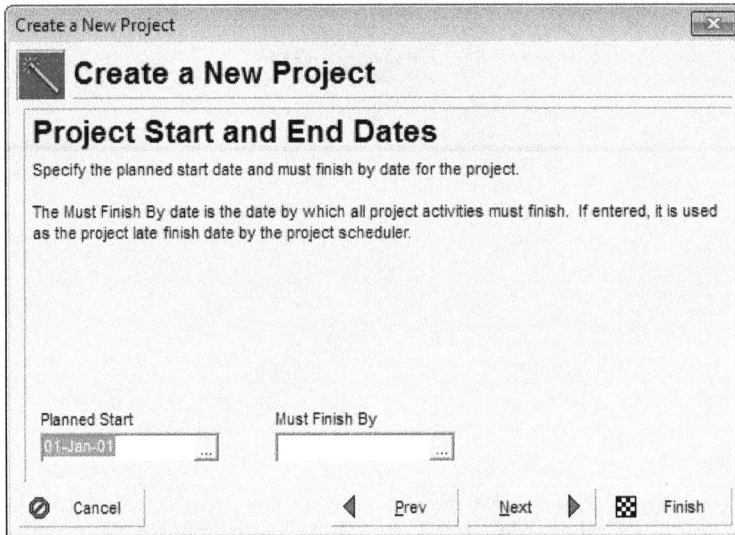

*Figure 2-6 The **Project Start and End Dates** page of the **Create a New Project** wizard*

The **Project Start and End Dates** page of the wizard is used to specify the starting date and desired ending date for the project. In the **Planned Start** edit box of this page you can specify the date on which you are planning to start your project. To specify a date, choose the Browse button in the **Planned Start** edit box; the calendar to specify the date will be displayed. Select a date from the calendar and then choose the **Select** button; the calendar will be closed and the date is specified in the **Planned Start** edit box. Similarly, in the **Must Finish By** edit box you can specify the date on which you want your project to get finished. On specifying the **Must Finish By** date enables you to track the project delay.

On specifying the start and finish date in the **Project Start and End Dates** page, choose the **Next** button; the **Responsible Manager** page of the **Create a New Project** wizard will be displayed.

In this page, you can assign a responsible manager to the project which is an OBS element and is selected from the OBS structure. To specify the responsible manager, click on the Browse button in the **Responsible Manager** edit box; the **Select Responsible Manager** dialog box will be displayed, as shown in Figure 2-7. In this dialog box, select the OBS name from the list that you want to assign as the responsible manager for your project. To do so, click on the desired name and then choose the **Select** button displayed on the right side of the **Select Responsible Manager** dialog box.

After specifying the responsible manager in the **Responsible Manager** page, choose the **Next** button; the **Assignment Rate Type** page of the **Create a New Project** wizard will be displayed, as shown in Figure 2-8.

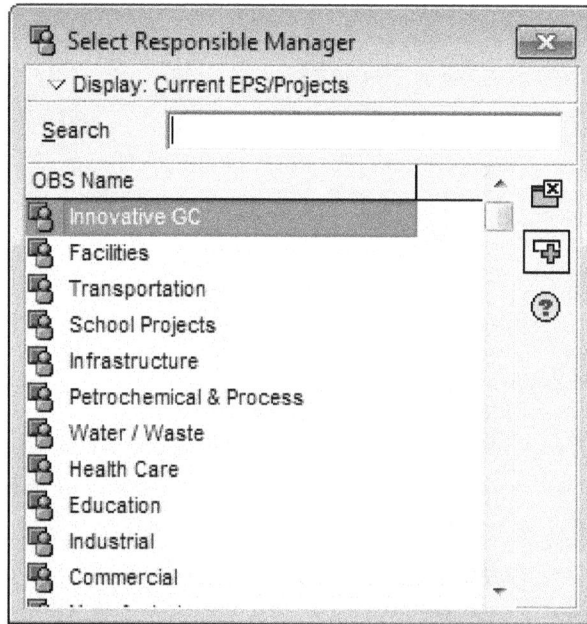

Figure 2-7 *The Select Responsible Manager dialog box*

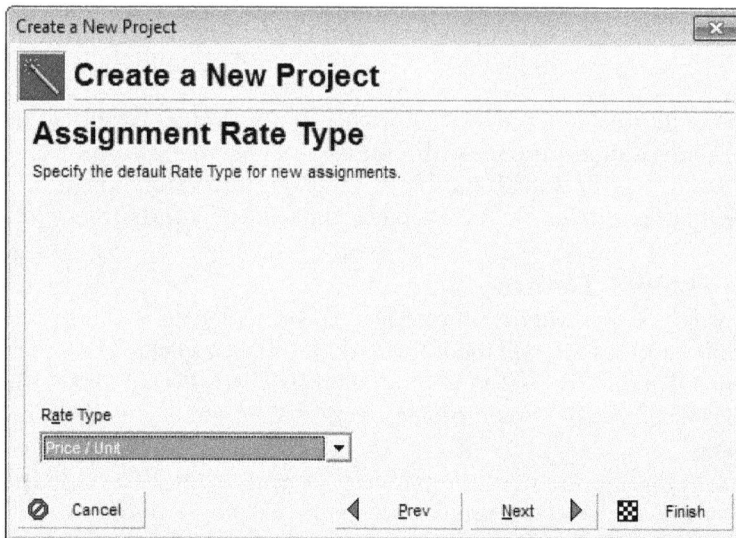

Figure 2-8 *The Assignment Rate Type page of the Create a New Project wizard*

The **Assignment Rate Type** page is used to specify the rate type you want to use to calculate costs for the assignments in your project. To specify the rate type, choose the desired price per unit from the **Rate Type** drop-down list. On specifying the rate type in the **Assignment Rate Type** page, choose the **Next** button; the **Project Architect** page of the **Create a New Project** wizard will be displayed, as shown in Figure 2-9.

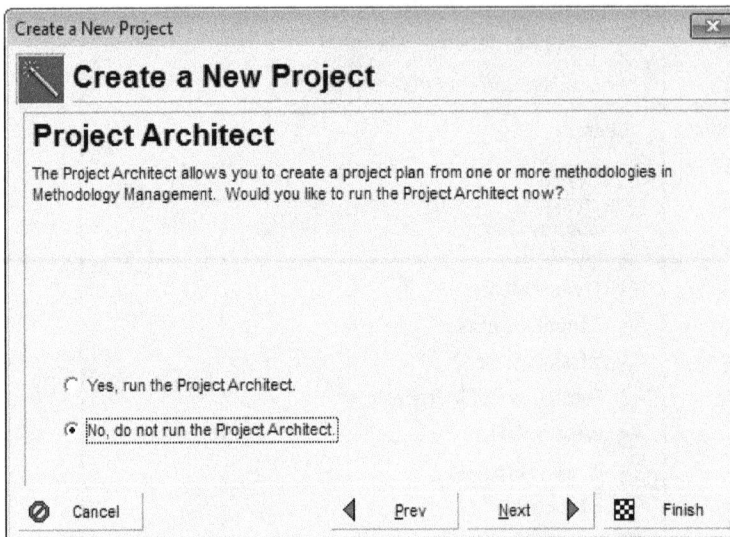

*Figure 2-9 The **Project Architect** page of the **Create a New Project** wizard*

This page is used to specify whether, you want to run the Project Architect to create a project plan from the methodologies available in the Methodology Management. The **No, do not run the Project Architect** radio button is selected by default. To run the project architect select the **Yes, run the Project Architect** radio button in the **Project Architect** page of the wizard and then choose the **Next** button; the **Welcome** page will be displayed. This page will allow you to create a new project by retrieving methodology content from the Methodology Management.

After specifying the desired settings for the project in the **Project Architect** page, choose the **Next** button; the **Congratulations** page with a message saying that your project has been created will be displayed. Choose the **Finish** button in the **Congratulations** page; the **Create a New Project** wizard will be closed and the new project will be added under the selected EPS node.

Defining the Project Status

The status of a project shows whether the project is completed or is still in progress. This will help you to organize and summarize the information. If the project is a completed project then it will be marked active and when it gets completed its status is changed to Inactive. You can also assign a What-if status to a copied project.

To assign a status to a project choose the required project or the EPS node from the **Projects** window; the details of the selected project will be displayed in different tabs of the **Project Details** table. In the **Project Details** table, choose the **General** tab and select the desired status for your project from the **Status** drop-down list, as shown in Figure 2-10. The **Active** status is selected by default in this drop-down list.

If you want the status of the project other than the **Active** status then place the project under separate root node in the EPS. Note that the project will still remain part of the hierarchy but will not be considered as the part of that EPS when you budget, schedule, and level your active projects. To place a project under separate root node, choose the **Enterprise Project Structure** option from the **Enterprise** menu in the menubar; the **Enterprise Project Structure (EPS)** dialog box will be displayed. In this dialog box, place the desired status root node at the bottom of the EPS to keep it separated from the rest of the hierarchy. To designate the separated node as the root node move the node to the left in the **Projects** window by using the **Shift Left** arrow button.

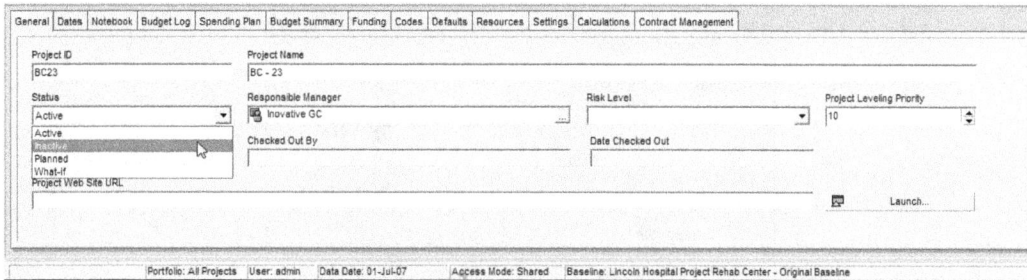

*Figure 2-10 Selecting the status from the **Status** drop-down in the **General** tab*

You can view projects with specific status. To do so, choose **Filter By > Customize** option from the **View** menu; the **Filters** dialog box will be displayed, as shown in Figure 2-11. In the **Filters** dialog box, select the desired check boxes corresponding to the desired status. Choose the **OK** button; the projects with the specified status will be displayed in the **Projects** window.

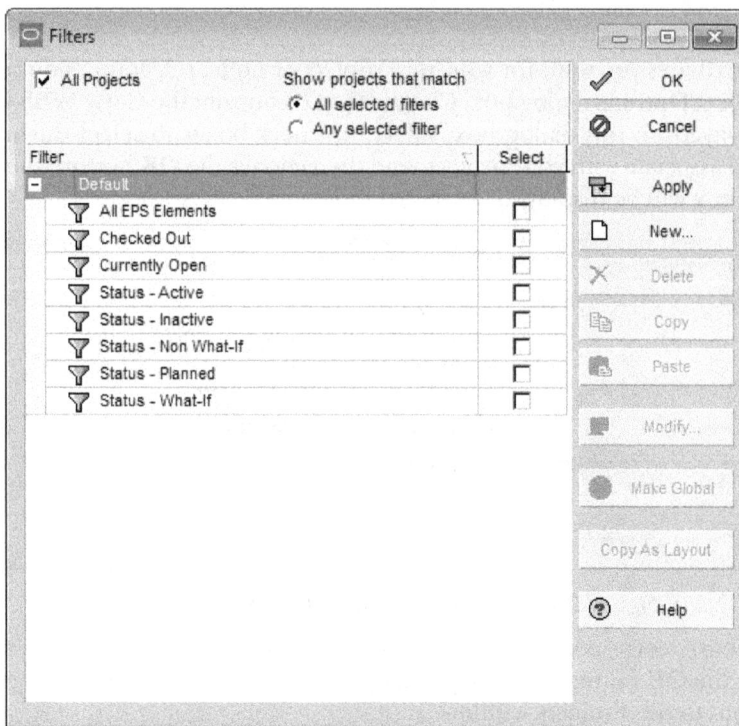

*Figure 2-11 The **Filters** dialog box*

Copying Project

You can copy an existing node or project by using the template of that node or project. The copied node or project then can be used as a template. When you copy a project in the EPS, you can also copy the related links to the WBS documents, OBS, and other related elements. To start with the copying process of a node or project, select the node or project you want to copy to the **Projects** window and then choose the **Copy** option from the Command bar. Next, choose the **Paste** option from the Command bar; the **Copy Project Options** dialog box will be displayed, as shown in Figure 2-12.

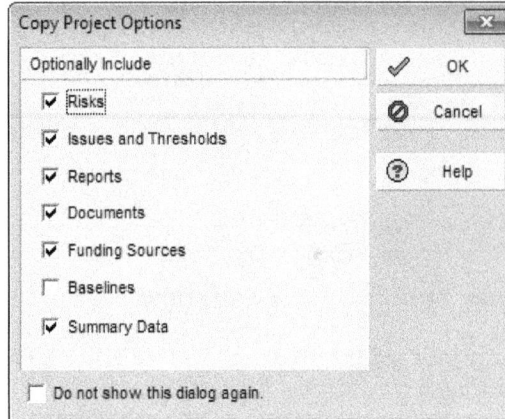

*Figure 2-12   The **Copy Project Options** dialog box*

To specify the attributes you want for your new project or node, select the required check boxes in the **Copy Project Options** dialog box. Choose the **OK** button; the **Copy WBS Options** dialog box will be displayed. In this dialog box, select the check boxes to select the attributes of the WBS you want to associate with your project, and then choose the **OK** button; the **Copy Activity Options** dialog box will be displayed, as shown in Figure 2-13.

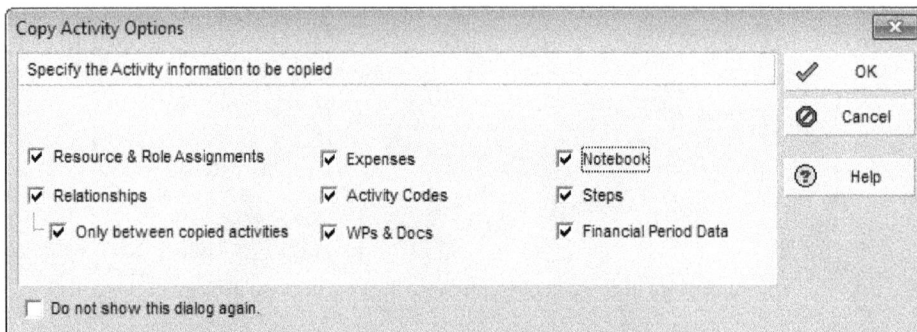

*Figure 2-13   The **Copy Activity Options** dialog box*

Select the check boxes corresponding to the desired option(s) in the **Copy Activity Options** dialog box and choose the **OK** button to exit the dialog box; the selected EPS node or project will be copied and added to the **Projects** window.

Deleting Project

You can also delete an EPS node or project. To do so, select the EPS node or project you want to delete and then choose the **Delete** option from the Command bar; the **Primavera P6** message box with a warning will be displayed, as shown in Figure 2-14.

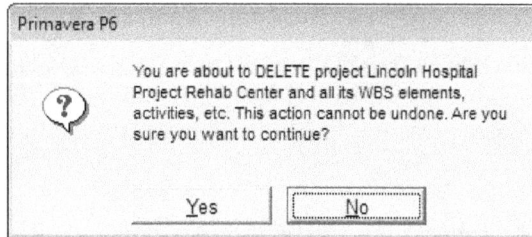

*Figure 2-14 The **Primavera P6** message box displayed*

This message box warns that if you delete this node or the project then all its related data which include the **WBS** elements, activities, and so on will also get deleted and once deleted cannot be recovered. If you choose the **Yes** button in this message box, the selected EPS node or project will be deleted.

> **Note**
> *When you delete an EPS node, all the projects under that branch of hierarchy will also be removed. So, if you do not want to delete those projects, copy and paste them to some other area in the hierarchy or under some other EPS node.*

Understanding Project Details Table

The **Project Details** table, as shown in Figure 2-15, is displayed under the **Projects** window. If not displayed by default, choose **View> Show on Bottom > Project Details** from the menubar.

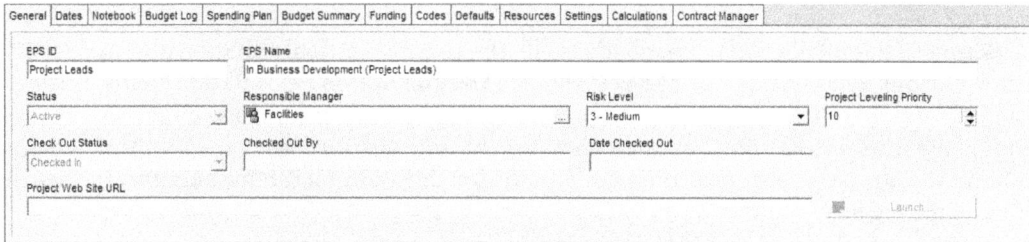

*Figure 2-15   The **Project Details** table*

The **Project Details** table contains various tabs having options to define an EPS or a project. You can customize the details table by right clicking anywhere in the table; a menu will be displayed. In this menu, select the **Customize Project Details** option; the **Project Details** dialog box will be displayed. In this dialog box, you can shift the required options from the **Available Tabs** list to the **Display Tabs** list. The tabs in the **Project Details** table are discussed next.

General

The **General** tab helps you to view and edit general information such as ID, name, status, responsible manager, risk level, project leveling priority for the selected node or project. If applicable, you can enter the project web site address in the **Project Web Site URL** edit box.

Dates

The **Dates** tab helps you to edit time table information for the selected project or node. This information includes the planned start date, must finish date, data date and finish date. Also, you can specify the anticipated start and finish date for the selected project.

Notebook

The **Notebook** tab helps you in assigning notebook topics, details, and description to the selected node or project.

Defaults

The **Defaults** tab enables you to specify the default settings or information for the selected project. This information includes the default cost account for resource assignments to activities, default automatic activity numbering, default activity calendar, duration type, and percent complete type.

Resources

The **Resources** tab helps you to specify resources essential for the project and project level resource permission for the time sheet application. These permissions include allowing resources to assign themselves to activities and to report their activities and assignments when completed. The options in the **Specify the default Rate Type for new assignments** drop-down list of this tab are used to calculate cost for activities that have labor/non-labor units with no assigned resources/ roles and do not have prices.

Settings

The **Settings** tab enables you to view and specify the summarized information and project-level settings for the selected node or project. You can use the options in different fields of this tab to summarize data for an EPS or a project and to automatically perform summary calculations using summary services. This tab also includes the settings for the first month of the fiscal calendar, the character for separating code fields for the WBS hierarchy, and also the settings to define the critical activities.

Calculations

The **Calculations** tab enables you to calculate the cost and estimate usage of resources when the activities in the selected project are updated. In this tab, the **Resource Assignments** area has two radio buttons for keeping track of your budget. If you select the **Add Actual to Remaining** radio button, then the amount will be at true complete units/costs. If you want to keep track of the amount remaining before you exceed your budget, select the **Subtract Actual from At Completion** radio button.

Select the **Recalculate Actual Units and Cost when duration % complete changes** check box to automatically update the actual units and costs when the duration % complete is updated. To recalculate units when costs are updated for resource assignments, select the **Update units when costs change on resource assignments** check box. If you want to store past period actuals in the **Store Period Performance** dialog box, then select the **Link Actual and Actual This Period Units and Cost** check box.

WORKING WITH THE OBS

The OBS, Organizational Breakdown Structure, is a hierarchical division of different project responsibilities between different managers in an organization. The OBS reflects the management structure of an organization from top-level personnel down, as shown in the Figure 2-16.

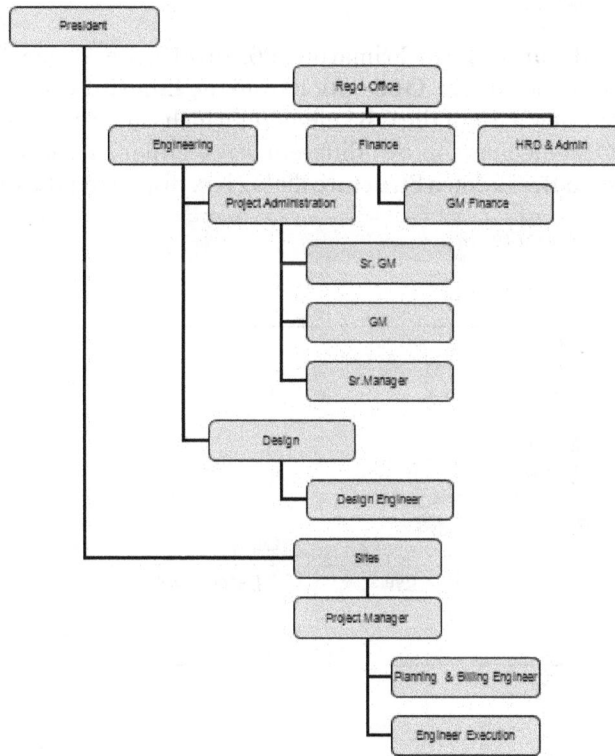

Figure 2-16 *The flowchart displaying the hierarchy of a typical organization setup*

Before starting a project, you need to identify the tasks included in a project and the individuals or team who will carry out those tasks. The OBS is an intermediate step in this process, it helps in identifying the employees who will be responsible for the project work.

In Primavera P6, the OBS elements are associated with the EPS nodes and projects. Access and privileges are assigned to the WBS/Projects/ EPS nodes through the OBS. Typically, an OBS is set up with a project manager at the project level to oversee, coordinate and manage the entire project. You can associate the responsible managers with their areas of the EPS in either nodes or individual projects. A responsible manager associated to an EPS node by default gets associated to the projects under the branch of the EPS.

Note
In Primavera P6, an OBS can have only one root element.

Adding OBS

Menu Bar: Enterprise > OBS

To create, view, and edit an OBS in Primavera P6, invoke the **Organizational Breakdown Structure** dialog box by choosing the **OBS** option from the **Enterprise** menu in the menubar. In this dialog box, you can view a list of the OBS elements which can be used in any desired project. You can display the OBS elements in two different ways, either by chart or table. Figure 2-17 shows the **Organizational Breakdown Structure** dialog box displaying the OBS in a tabular form.

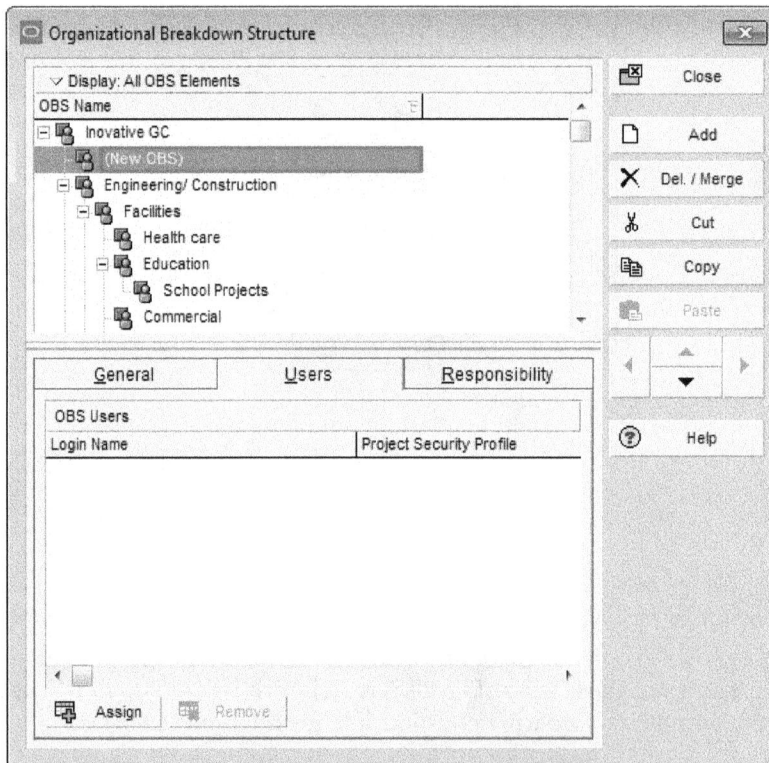

*Figure 2-17   The **Organizational Breakdown Structure** dialog box*

You can create a new OBS by selecting the existing OBS element or node immediately above or at the same hierarchy level. To add a new element, choose the **Add** button; the new OBS element will be added under and at the same hierarchy level as of the selected OBS node.

You can specify all the information related to the created OBS by using the options available in the **General**, **Users,** and **Responsibility** tabs of the **Organizational Breakdown Structure** dialog box. These tabs are discussed next.

General

In the **General** tab, you can specify the name of the created OBS in the **OBS Name** edit box and if required you can provide description of the OBS in the **OBS Description** edit box. The editing tools displayed under the **OBS Description** edit box are used to format text, insert pictures, copy, and paste information from other document files, and to add hyper links.

Users

In the **Users** tab, you can view the users and corresponding security profiles affiliated with the OBS element. Also, if you have proper access rights then you can assign users to the OBS.

Responsibility

In this tab, the **Project ID/WBS Code** column lists the project ID/WBS code for which the selected OBS element is responsible. The **Project Name/WBS Name** column lists the project name and WBS name for which the selected OBS element is responsible.

You can also copy, paste, and delete an OBS. To do so, select the desired OBS element, choose the **Copy** button; the desired OBS will be copied and then select the OBS to which you want to add the copied OBS element and choose the **Paste** button. The copied OBS element will be added to the desired OBS. Similarly, you can cut and paste an OBS element.

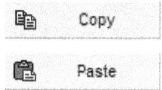

You can also delete an OBS element. To do so, select the required OBS and choose the **Del. /Merge** button from right pane of the **Organizational Breakdown Structure** dialog box; the selected OBS element will be deleted.

TUTORIALS

Tutorial 1	**Creating EPS and OBS**

In this tutorial, you will create EPS and OBS for a project. **(Expected time: 30 min)**

The following steps are required to complete this tutorial:

a. Open the **Projects** window.
b. Create EPS.
c. Assign OBS.
d. Create a new project.

Opening the Projects Window

1. Open Primavera P6 and invoke the **Projects** window by choosing the **Projects** option from the **Enterprise** menu in the menubar.

Creating EPS

1. Choose the **Enterprise Project Structure** option from the **Enterprise** menu in the menubar; the **Enterprise Project Structure (EPS)** dialog box is displayed.

2. In the **Enterprise Project Structure (EPS)** dialog box, select the **Innovative Construct** EPS node under which the new EPS element is to be added.

3. Next, choose the **Add** button from the right pane of the **Enterprise Project Structure (EPS)** dialog box; a new EPS with the name **NEWEPS** is added to the **Innovative Construct** EPS node, as shown in Figure 2-18.

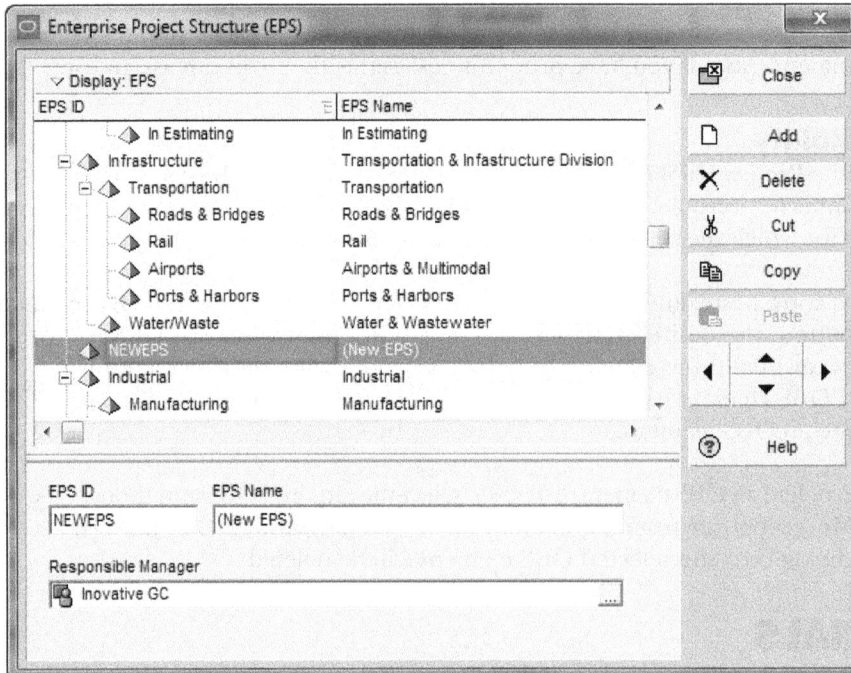

Figure 2-18 A new EPS added under the Innovative Construct EPS node in the Enterprise Project Structure (EPS) dialog box

4. Click on **NEWEPS** in the **EPS ID** column and rename it to **CADCIM**. Change the name of the **EPS** in the **EPS Name** column to **CADCIM Technologies** and press ENTER.

5. Ensure that the **CADCIM** EPS is selected and then choose the **Shift Left** button from the Command bar; the **CADCIM** EPS node is now an independent node.

6. Next, choose the **Shift Up** button from the Command bar; the node is shifted on top, as shown in Figure 2-19.

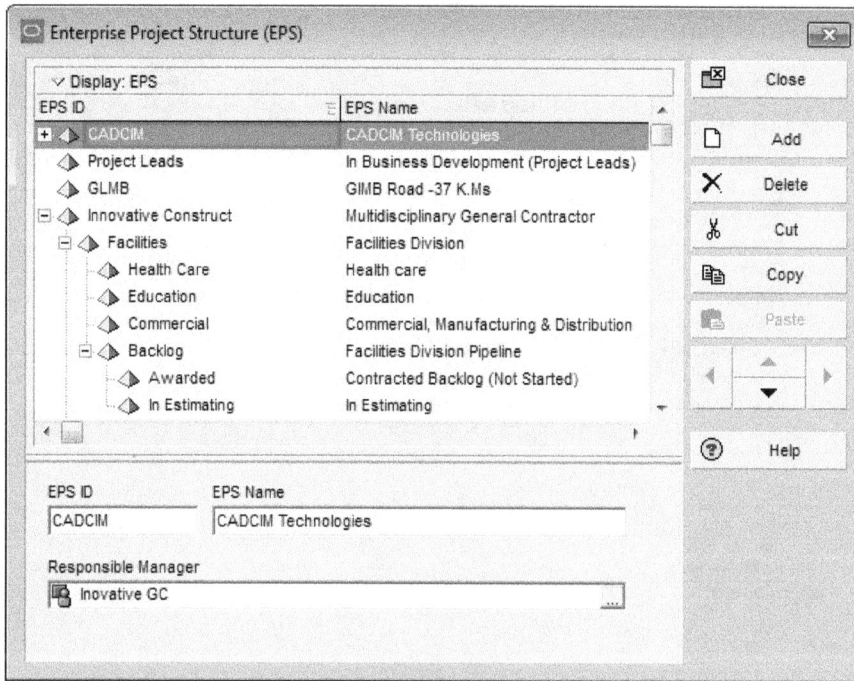

*Figure 2-19 The **CADCIM** EPS node displayed at the top*

7. Ensure that the **CADCIM** EPS is selected and then choose the **Add** button; a new EPS node is added under the **CADCIM** EPS.

8. Double-click on the **NEWEPS** node in the **EPS ID** column and change the ID of the created EPS to **Construction** and **EPS Name** to **Construction Division**.

9. Similarly, add two more EPS under the **CADCIM** head with the **EPS ID** as **Publishing**, and **Consultancy** respectively and the **EPS Name** as **Publishing Division**, and **Consultancy Division** respectively, refer to Figure 2-20.

10. Choose the **Close** button from the right pane of the dialog box to close the **Enterprise Project Structure (EPS)** dialog box.

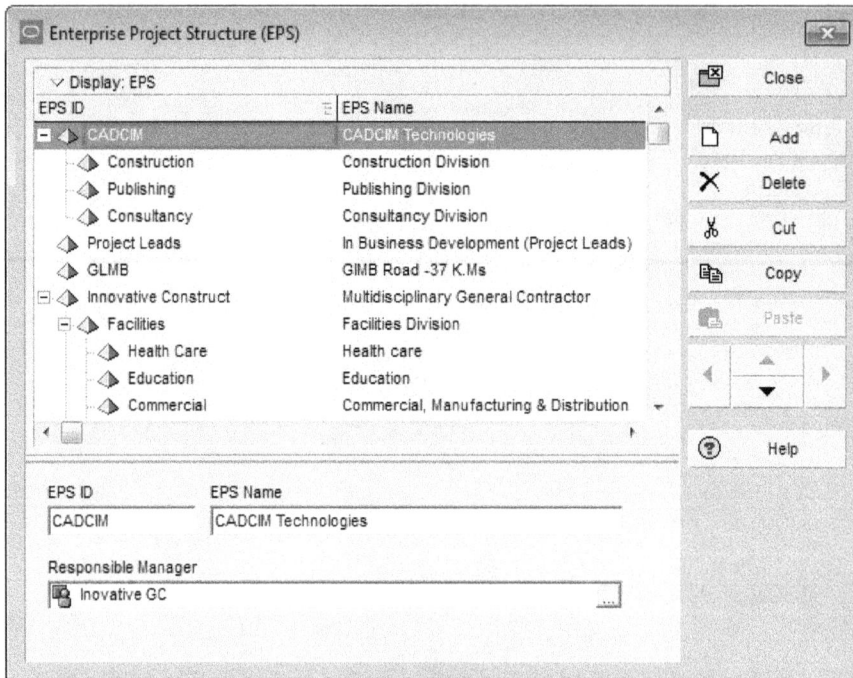

Figure 2-20 *The **CADCIM** EPS node with further subnodes displayed*

Creating OBS

1. Choose the **OBS** option from the **Enterprise** menu in the menubar; the **Organizational Breakdown Structure** dialog box is displayed.

2. Select the **Filter By > All OBS Elements** option from the **Display** drop-down list; all the OBS elements are displayed in the dialog box.

3. In this dialog box, select the **Construction-Tim Evans** option and then select the **Add** button from the Command bar, a new OBS with the name **NEWOBS** is created.

4. Choose the **Shift Up** button to shift the newly created OBS to the top.

5. Rename the created OBS as **Company Head- Sham Tickoo** in the **OBS Name** edit box.

6. Choose the **Add** button, a new OBS is created. Choose the **Shift Right** button to shift the OBS to the right.

7. Add two more OBS under Company Head and ensure that all the added OBS are shifted to the right.

8. Rename the added OBS as **Project Manager**, **Publishing Head**, and **Consultants Head**, as shown in Figure 2-21.

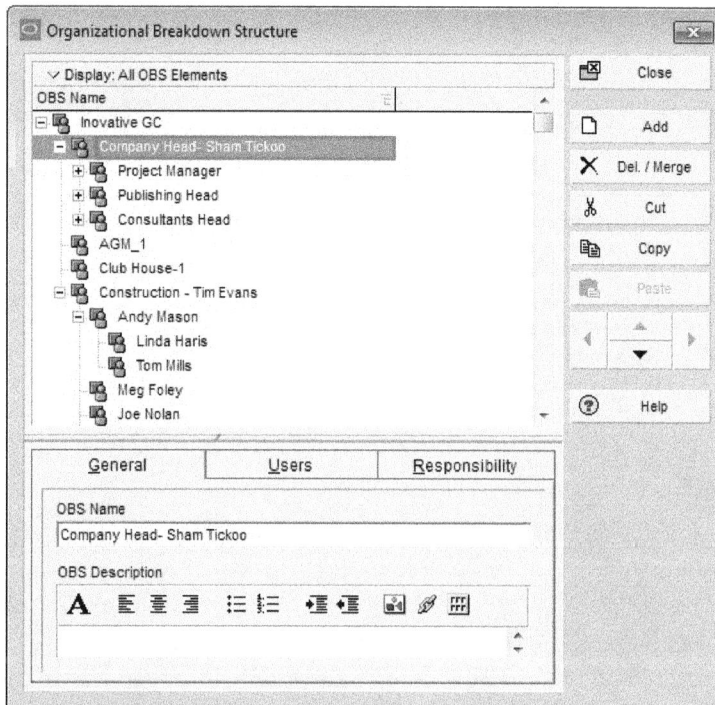

Figure 2-21 Organizational Breakdown Structure displayed with the structured OBS elements

9. Similarly, add more OBS in the order, as shown in Figure 2-22.

10. Next, choose the **Close** button; the **Organizational Breakdown Structure** dialog box is closed and the **Projects** window is displayed.

11. Ensure that the **Construction Division** node is selected in the **Projects** window.

12. Now, choose the browse button adjacent to the **Responsible Manager** edit box in the **General** tab of the **Project Details** table; the **Select Responsible Manager** dialog box is displayed.

13. In this dialog box, select the **Company Head- Sham Tickoo** node from the OBS Name list.

14. Choose the **Select** button from the Command bar; the responsible manager is assigned to the construction division.

15. Similarly, assign **Company Head-Sham Tickoo** as the OBS to the **Publishing** and **Consultancy** division node.

The EPS and OBS are created and are saved in the Primavera P6.

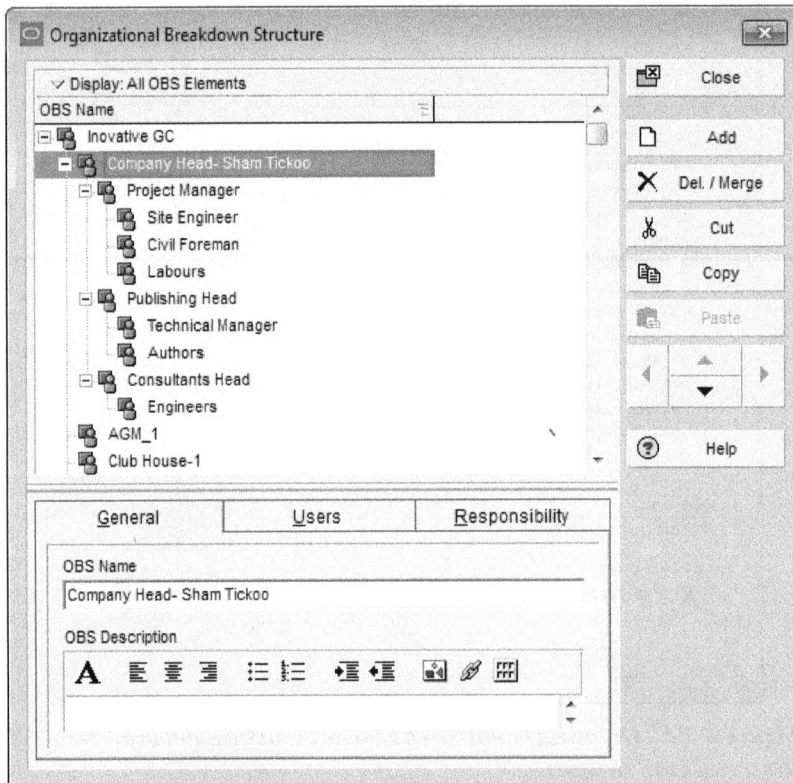

*Figure 2-22 The **Organizational Breakdown Structure** is displayed with the complete structured OBS*

Tutorial 2   Creating a Project

In this tutorial, you will create a project and will assign the OBS to the project. Remember to complete tutorial 1 before this.   **(Expected time: 45 min)**

The following steps are required to complete this tutorial:

a.   Open the **Projects** window.
b.   Create a Project
c.   Export a Project.

Opening the Projects Window

1.   Open Primavera P6 and invoke the **Projects** window by choosing the **Projects** option from the **Enterprise** menu in the menubar.

Creating a Project

1.   In the **Projects** window, select the **Construction** EPS node from the **CADCIM** head. Choose the **Add** button from the Command bar; the **Create a New Project** wizard with the **Select EPS** page is displayed, as shown in the Figure 2-23.

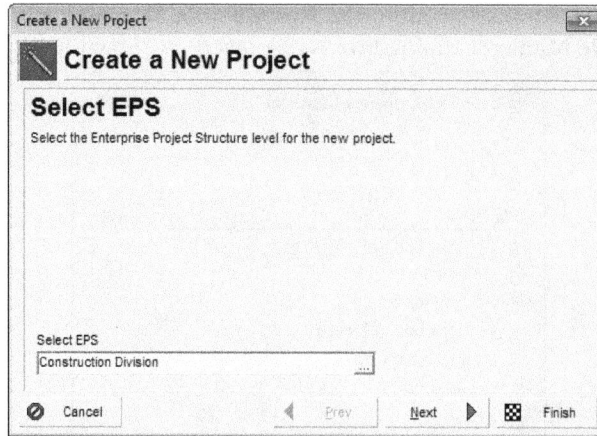

*Figure 2-23 The **Create a New Project** wizard with the **Select EPS** page*

2. In this **Select EPS** page, the **Construction Division** EPS is chosen by default. Keep the default settings in this page and choose the **Next** button; the **Project Name** page of the **Create a New Project** wizard is displayed.

3. In this page, specify **CD00720** as the project id and **Home Construction** as the project name in the **Project ID** and **Project Name** edit boxes, respectively.

4. Choose the **Next** button; the **Project Start and End Dates** page is displayed.

5. In the **Project Start and End Dates** page, choose the browse button located adjacent to the **Planned Start** edit box; a flyout calendar is displayed.

6. Using the flyout calendar, set the start date for the project to **13-Oct-15** and choose the **Select** button from the flyout.

7. Choose the **Next** button; the **Responsible Manager** page is displayed, as shown in Figure 2-24.

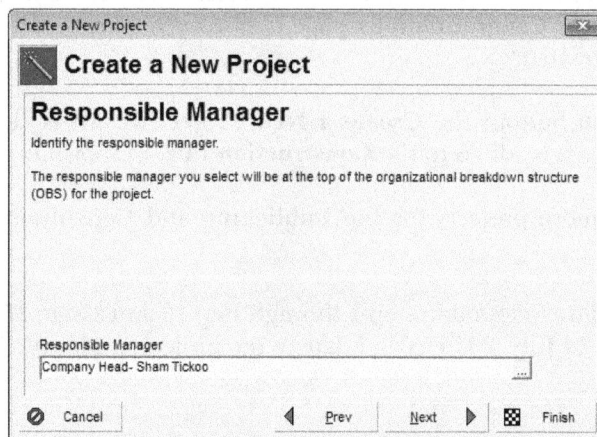

*Figure 2-24 The **Responsible Manager** page of the **Create a New Project** wizard*

8. Choose the browse button adjacent to the **Responsible Manager** edit box of this page; the **Select Responsible Manager** dialog box is displayed, as shown in Figure 2-25.

Figure 2-25   The Select Responsible Manager dialog box

9. If the elements are not displayed in this dialog box by default, select **Filter By > All OBS Elements** option from the **Display** drop-down list.

10. Select the **Project Manager** option in the **OBS Name** list and choose the **Select** button from the right pane of the dialog box.

11. Choose the **Next** button in the **Responsible Manager** page of the **Create a New Project** wizard; the **Assignment Rate Type** page is displayed.

12. Retain all the default settings in this page and choose the **Next** button; the **Project Architect** page is displayed.

13. In this page, select the **No, do not run the Project Architect** radio button and then choose the **Next** button; the **Congratulations** page is displayed with the message that your new project has been created.

14. Choose the **Finish** button; the **Create a New Project** wizard is closed and the **Home Construction** project is added to the **Construction** EPS, as shown in the Figure 2-26.

    Similarly, create more projects for the **Publishing** and **Consultancy** Division, refer to Figure 2-27.

15. Repeat the procedure followed in step 1 through step 13 and assign **15 December, 2014** as the start date and **24 July, 2015** as end date to the projects under the **Publishing Division** node.

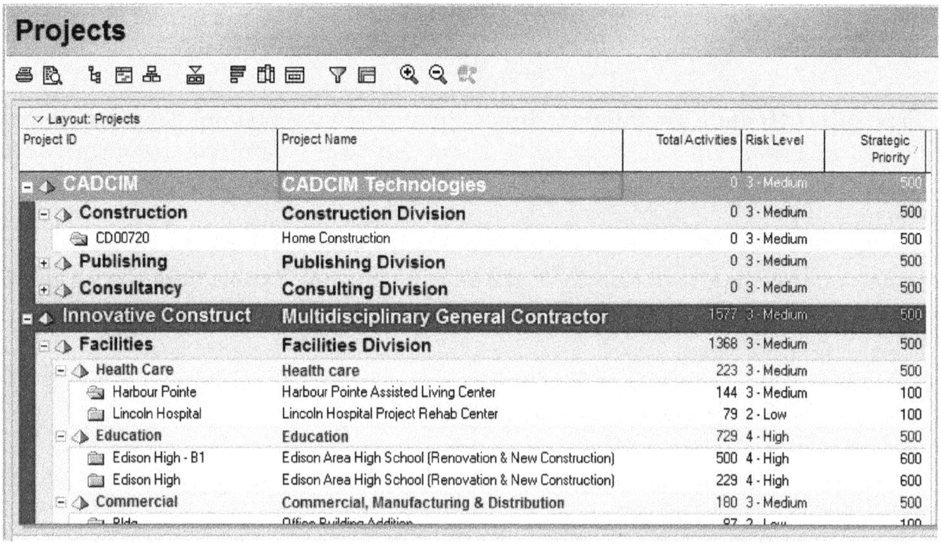

*Figure 2-26 The **Home Construction** project added under the **CADCIM** EPS*

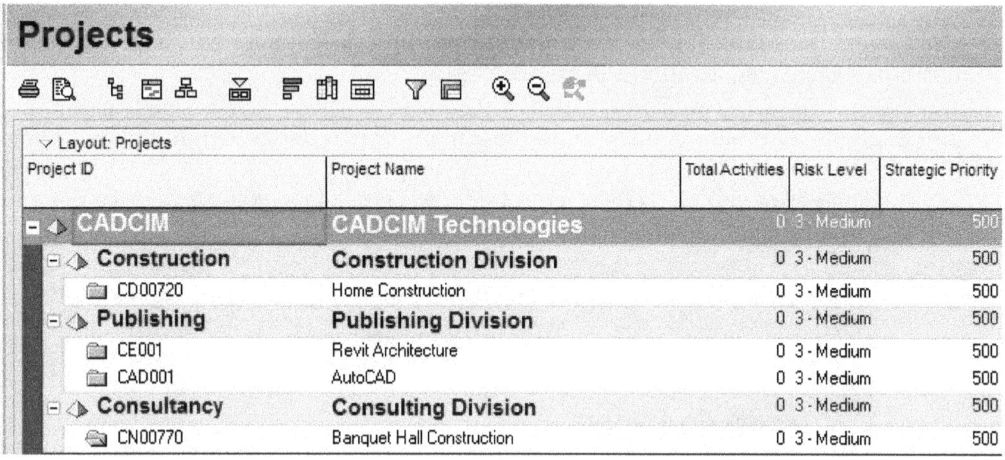

*Figure 2-27 All projects added under the **CADCIM** EPS*

16. For **Consultancy Division** project, enter **5th November, 2014** in the **Planned Start** edit box.

17. In the **Responsible Manager** page, assign the **Author** as **Responsible Manager** for the **Publishing Division** and **Engineer** for the **Consultancy Division**.

Primavera P6 automatically saves the created EPS and project.

Exporting Projects

Export projects to save the project file for future reference.

1. To export the **Home Construction** project, ensure that the **Home Construction** project is opened in Primavera P6. If not, then select the **Home Construction** project from the **Projects** window and right-click; a menu is displayed. Choose the **Open Project** option from the menu.

2. Now, choose the **Export** option from the **File** menu; the **Export** wizard with the **Export Format** page is displayed.

3. In this page, ensure that the **Primavera PM/MM - (XER)** radio button is selected and then choose the **Next** button; the **Export Type** page is displayed.

4. In this page, ensure that the **Project** radio button is selected and then choose the **Next** button; the **Projects To Export** page is displayed.

5. Make sure the **Home Construction** project is displayed in this page, as shown in Figure 2-28. Choose the **Next** button; the **File Name** page is displayed.

6. In this page, choose the Browse button adjacent to the **File Name** edit box; the **Save File** dialog box is displayed, as shown in Figure 2-29.

7. In this dialog box, browse to the C drive and then create a folder with the name *PM6*.

8. Open the folder *PM6* and then create a sub-folder with the name **c02**. Next, open the created folder and save the file with the name *c02_CONS_Home_tut02* and then choose the *Save* button.

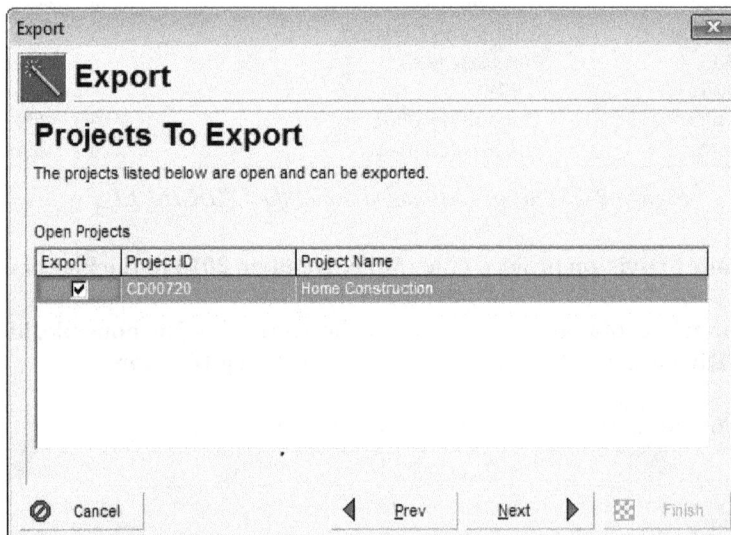

Figure 2-28 *The* **Home Construction** *project displayed in the* **Projects To Export** *page*

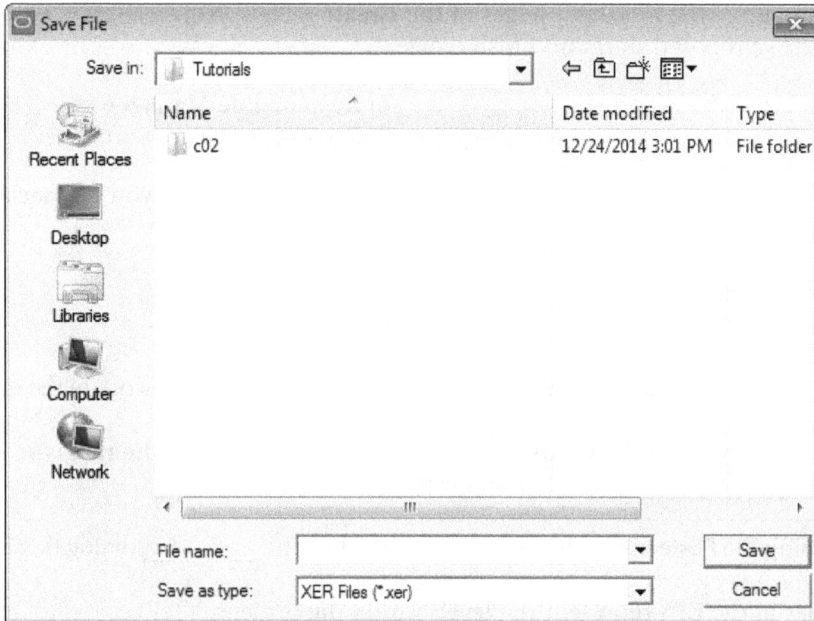

Figure 2-29 The Save File dialog box

9. Choose the **Finish** button; the **Primavera P6** message box is displayed showing the message that the export was successful.

10. Choose the **OK** button; the file is exported.

11. Similarly, export other projects and save them at the same location with the name as mentioned below.
    AutoCAD- **c02_PUB_CAD_tut02**
    Revit Architecture- **c02_PUB_Archi_tut02**
    Banquet Hall Construction- **c02_CN_BNQT_tut02**

Self-Evaluation Test

Answer the following questions and then compare them to those given at the end of this chapter:

1. In which of the following tabs of the **Organizational Breakdown Structure** dialog box, you can specify the name and description of the OBS?

    a) **General**   b) **Users**
    c) **Responsibility**   d) none of the above

2. In the **Project Details** table, which of the following tabs enables you to view and specify the summarization information and project-level settings for the selected node or project?

    a) **Resources**   b) **Notebook**
    c) **Settings**   c) **Calculations**

3. In which page of the following pages of the **Create a New Project** wizard, you can specify the project's proposed start and finish dates?

   a) **Select EPS**   b) **Responsible Manager**
   c) **Project Name**   d) **Project Start and End Dates**

4. In the **Project Details** table, which of the following tabs enables you to specify resources essential for the project?

   a) **Notebook**   b) **Budget Log**
   c) **Codes**   d) **Resources**

5. An **EPS** can further be sub-divided into many_____ according to work in the organization.

6. The _____ page of the **Create a New Project** wizard specifies the rate type you want to use to calculate costs for the assignments in your project.

7. On choosing the **Paste** button from the Command bar, the _____ dialog box is displayed.

8. The nodes in the EPS represent the levels within the projects. (T/F)

9. In Primavera P6, access and privileges to the WBS/Projects/ EPS nodes are assigned through the OBS. (T/F)

10. The status of a project is shown by marking it **Active** if it is not completed and **Inactive** when it gets completed. (T/F)

Review Questions

Answer the following questions:

1. In the **Project Details** window, which of the following tabs helps you in assigning notebook topics, details, and description to the selected node or project?

   a) **Calculations**   b) **Notebook**
   c) **Defaults**   d) **General**

2. You can invoke the **Organizational Breakdown Structure** dialog box by choosing the **OBS** option from the _____ menu in the menubar.

3. In the _____ page of the **Create a New Project** wizard, you can specify the **EPS** under which you want to add the project.

4. To invoke the **Enterprise Project Structure** dialog box, choose the _____ option from the **Enterprise** menu in the menubar.

5. The **EPS** nodes that contains subnodes and projects are identified by a _____ symbol.

6. You can use the _____ button from the Command bar of the **Enterprise Project Structure (EPS)** dialog box to indent/outdent a node to mark its placement in the **EPS** and to move a node up and down in the hierarchy.

7. The **Calculations** tab enables you to calculate cost and estimate resources to be used when you update activities in the selected project. (T/F)

8. You can add activities to your new project by using the methodologies available in **Methodology Management** module of Primavera P6 by running the project architect. (T/F)

9. When you delete an **EPS** node, all the projects under that branch of hierarchy will also be removed. (T/F)

10. You can use the template of an existing node or project to create a new one by copying the existing node or project. (T/F)

EXERCISES

Exercise 1

Create an Enterprise Project Structure (EPS), as shown in Figure 2-30 and Figure 2-31 and Organizational Breakdown Structure, as shown in Figure 2-32 for an enterprise of construction department. Assume the name of the enterprise as ABC Technologies.   **(Expected time: 30 min)**

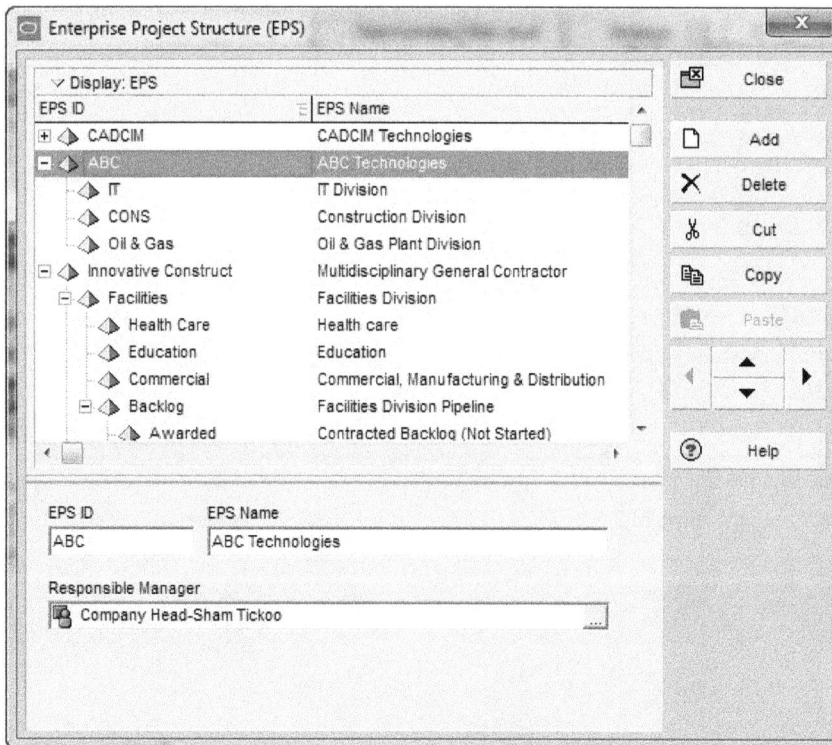

*Figure 2-30 The **Enterprise Project Structure (EPS)** dialog box with the ABC EPS*

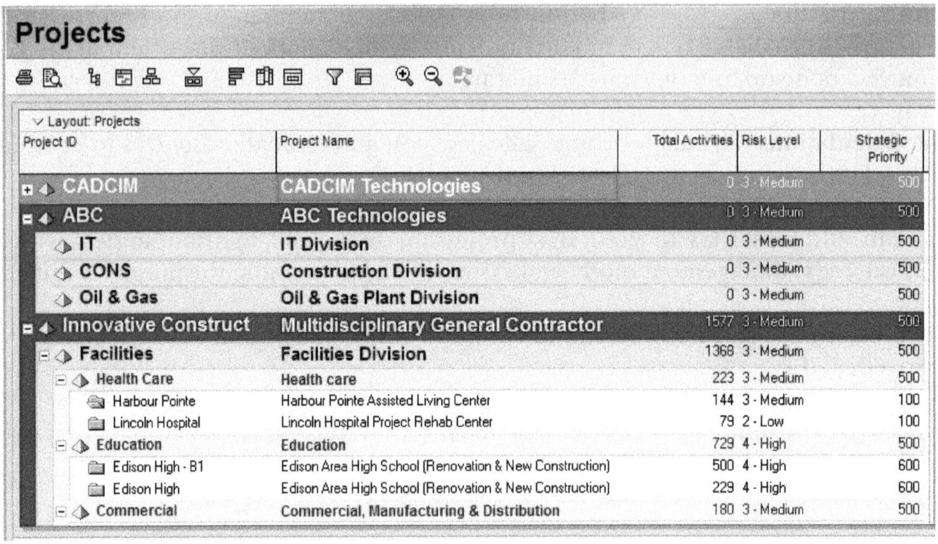

Figure 2-31 The EPS created in the Projects window

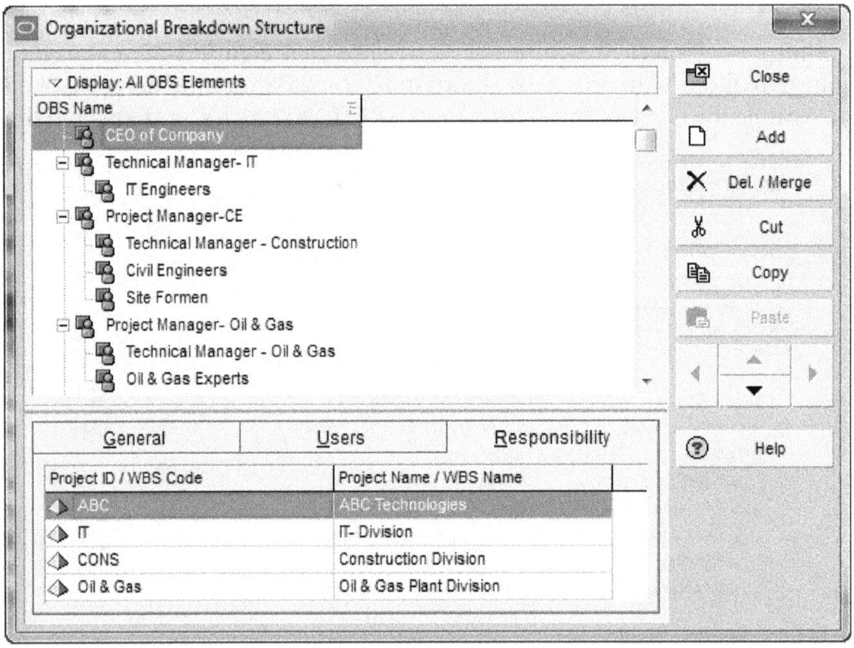

Figure 2-32 The OBS created for ABC Technologies project

Exercise 2

Create projects for the ABC Technologies under different EPS, as shown in Figure 2-33. You can download project files related to these projects from the *www.cadcim.com*.

(Expected time: 45 min)

1. **IT Division**
a. Project I   **Project Name**   Android Application Development
                 Project ID   AAD001
                 Project Start Date   22-Sep-15
                 Responsible Manager   IT Engineers
                 Save File   c02_IT_App_ex02

b. Project II   **Project Name**   Web Designer Java
                 Project ID   WD002
                 Project Start Date   25-Aug-15
                 Responsible Manager   IT Engineers
                 Save File   c02_IT_WEB_ex02

2. **Construction Division**
                 Project Name   Hospital Building Project
                 Project ID   CD00770
                 Project Start Date   05-Nov-15
                 Responsible Manager   Technical Manager - Construction
                 Save File   c02_CONS_HOSP_ex02

3. **Oil & Gas Power Plant**
a. Project I   **Project Name**   XYZ Oilfield project
                 Project ID   XYZ004
                 Project Start Date   15-Dec-15
                 Responsible Manager   Technical Manager- Oil & Gas
                 Save File   c02_O&G_XYZ_ex02

b. Project II   **Project Name**   CSZ Gasfield project
                 Project ID   CSZ005
                 Project Start Date   20-Oct-15
                 Responsible Manager   Technical Manager- Oil & Gas
                 Save File   c02_O&G_CSZ_ex02

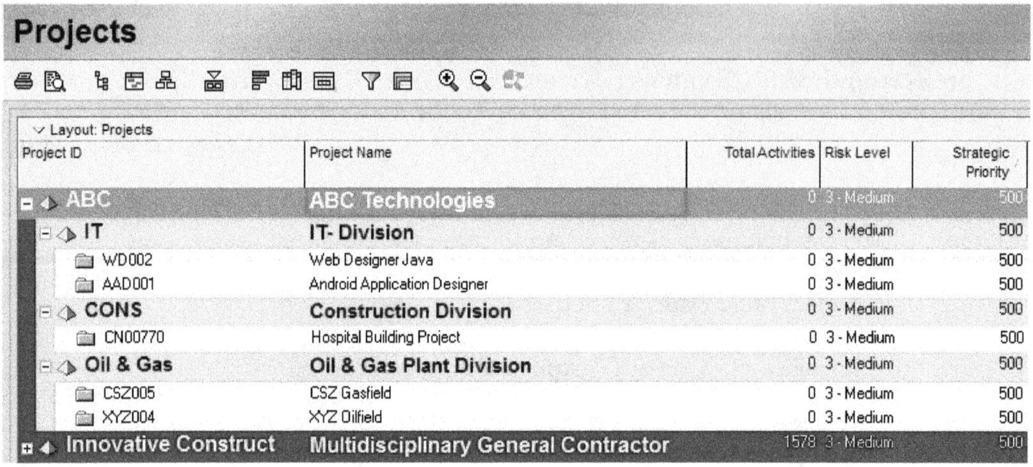

Projects

Project ID	Project Name	Total Activities	Risk Level	Strategic Priority
ABC	ABC Technologies	0	3 - Medium	500
IT	IT- Division	0	3 - Medium	500
WD002	Web Designer Java	0	3 - Medium	500
AAD001	Android Application Designer	0	3 - Medium	500
CONS	Construction Division	0	3 - Medium	500
CN00770	Hospital Building Project	0	3 - Medium	500
Oil & Gas	Oil & Gas Plant Division	0	3 - Medium	500
CSZ005	CSZ Gasfield	0	3 - Medium	500
XYZ004	XYZ Oilfield	0	3 - Medium	500
Innovative Construct	Multidisciplinary General Contractor	1578	3 - Medium	500

*Figure 2-33 The **Projects** created in the **Projects** window*

Answers to Self-Evaluation Test

1. a, 2. c, 3. d, 4. d, 5. nodes, 6. Assignment Rate Type, 7. Copy Project Options, 8. T, 9. T, 10. T

Chapter 3

Defining Calendars and Work Breakdown Structure

Learning Objectives

After completing this chapter, you will be able to:
- *Create a new calendar*
- *Change a project calendar into global calendar*
- *Modify a calendar*
- *Define WBS for a project*
- *Add WBS elements and assign properties*
- *Use WBS milestones*

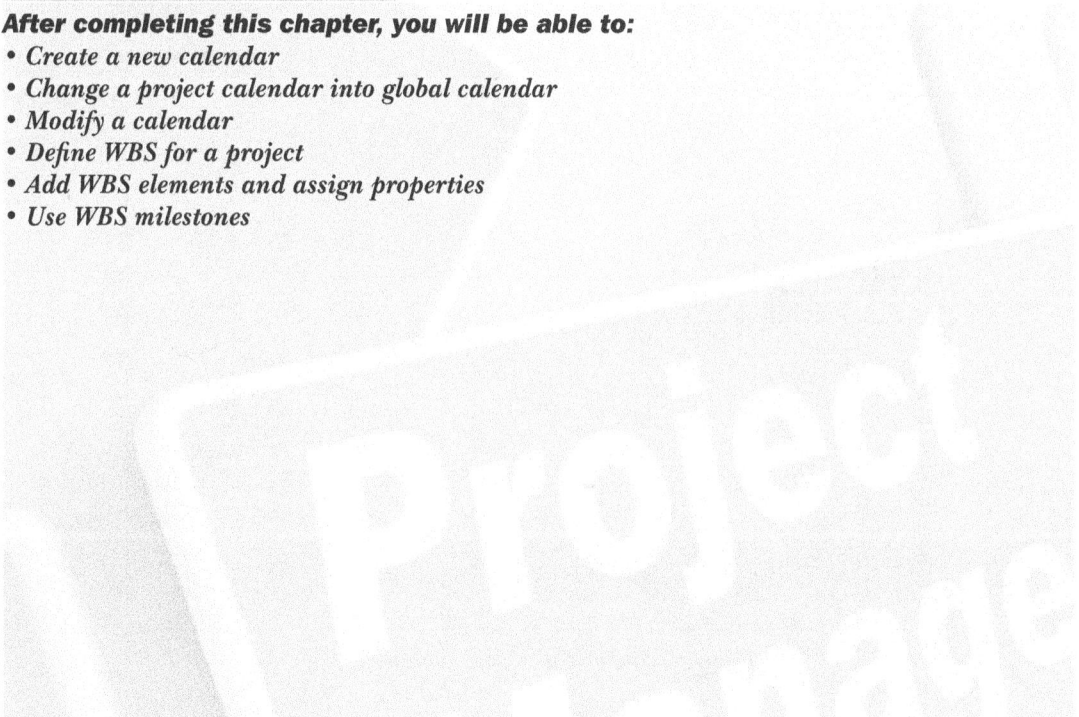

INTRODUCTION

In this chapter, you will learn about the calendars in Primavera P6 and will also learn different procedures to create them. The method of deleting, renaming, and modifying a calendar has also been explained in detail in this chapter.

This chapter also explains the procedure of defining WBS for a project, WBS properties, and the usage of WBS milestones. The process of adding activities to a project is also discussed in this chapter.

CREATING AND ASSIGNING CALENDAR

A calendar in Primavera P6 enables you to define the number of available workdays as well as workhours in a workday. You can also specify public holidays, week off of the organizations, and project specific working and non-working days.

You can create different types of calendars and assign them to specific resources and activities to determine the time constraints in a uniform way. By using calendars, you can do activity scheduling, tracking, and resource leveling. Primavera P6 consists of three types of calendars: **Global**, **Resource**, and **Project**. These calendar types are discussed next.

Global Calendar

Global calendars are those that can be assigned to multiple projects. If a calendar is defined under this category, then that calendar can be assigned to any project, activity, and resource in your database.

Creating a Global Calendar

To create a global calendar, choose the **Calendars** option from the **Enterprise** menu in the menubar; the **Calendars** dialog box will be displayed, refer to Figure 3-1.

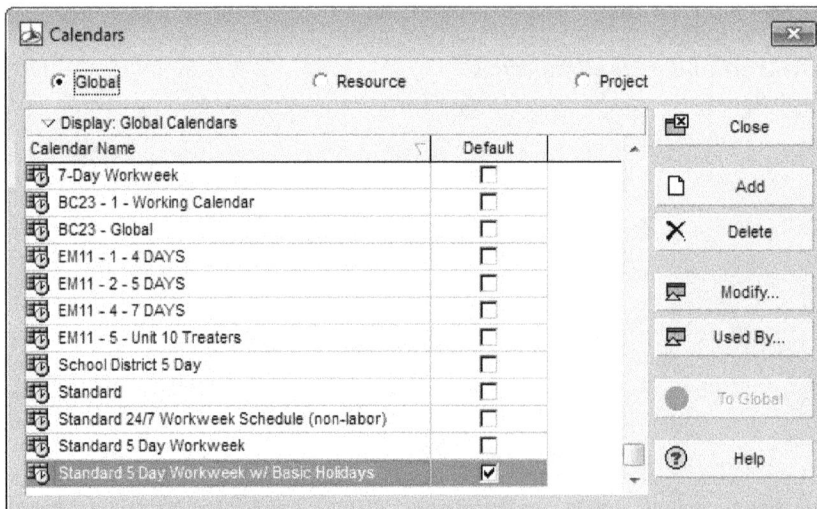

*Figure 3-1 The **Calendars** dialog box*

In this dialog box, select the **Global** radio button if not selected by default; a list of all the available global calendars will be displayed in the **Calendar Name** area. To create a new global calendar, choose the **Add** button from the right pane of the **Calendars** dialog box; the **Select Calendar To Copy From** dialog box will be displayed, as shown in Figure 3-2.

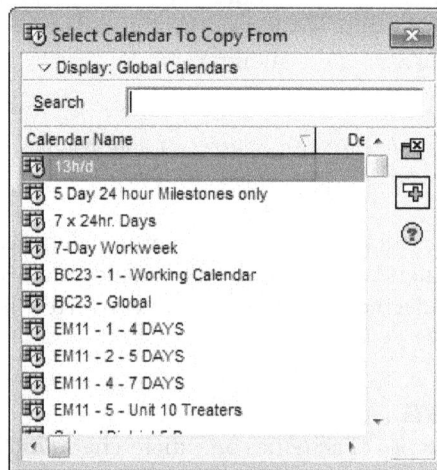

Figure 3-2 The Select Calendar To Copy From dialog box

Next, select the required calendar and then choose the **Select** button; a new calendar will be added and the **Calendar To Copy From** dialog box will be closed.

Project Calendar

Project calendars are calendars which can be assigned to some specific project. Calendars under this category can not be assigned to any resources such as labor, non-labor, and material. If you link a project calendar to a global calendar then all the holidays and the exceptions specified in the global calendar will reflect in the project calendar also but the changes made in the project calendar will not reflect in the global calendar.

Copying a Project Calendar from One Project to Another Project

You can copy a calendar from one project to another. To copy the project calendar of a project to any other project, the project calendar is to be converted into a global calendar.

Resource Calendar

As the name suggests the resource calendars are those calendars that are defined for resources only. The calendars defined under this category can only be assigned to specific resources and cannot be assigned to a project or its activities. If the resource calendar is linked to a global calendar than any change made in the global calendar will replace in the resource calendar but not vice-versa.

Creating a Project or Resource Calendar

Choose the **Calendars** option from the **Enterprise** menu in the menubar; the **Calendars** dialog box will be displayed, refer to Figure 3-1. In this dialog box, select the **Project** or **Resource** radio button depending upon the type of calendar to be created; a list of all the available calendars will be displayed under the **Calendar Name** area. Next, choose the **Add** button from the right pane of the **Calendars** dialog box; the **Select Calendar To Copy From** dialog box will be displayed.

In this dialog box, select the global calendar that you want to copy as the new calendar and choose the **Select** button. On doing so, the **Calendar To Copy From** dialog box will be closed and the newly created calendar will be added to the **Calendar** section of the desired calendar type. Assign a name to the newly created calendar.

> **Note**
> *The global and resource calendars can be accessed and modified even when no project is opened whereas the project calendar can be modified only when a project has been opened.*

A project or resource calendar can also be converted into a global calendar. To do so, select the required project or resource calendar in the **Calendars** dialog box. Next, choose the **To Global** button; the selected calendar will be converted into a global calendar.

Modifying Calendars

A project consists of different types of activities on which scheduling is done. Therefore, it is very difficult to create a single calendar that can be assigned to each activity for the same purpose. To cover each and every activity in a project, you may have to create different calendars with different workday and workhour schedules. For example, you can create a calendar with normal Monday through Friday workdays and another calendar which has continuous work time (24hours/day). Every activity in a project must be assigned to a specific calendar that specifies the work time available to perform that activity.

With so many calendars and activities changing with time, there is always a need to modify the existing calendars to match up with the new upcoming activities. The different methods for modifying the calendars are discussed next.

Modifying the Working Days in the Calendar

To modify the working days in a calendar, choose the **Calendars** option from the **Enterprise** menu of the menubar; the **Calendars** dialog box will be displayed. In this dialog box, select the calendar type (**Global**, **Project**, or **Resource**) and then select the calendar that you want to modify. Next, choose the **Modify** button; the dialog box corresponding to the selected options will be displayed, refer to Figure 3-3. In this dialog box, you can change the month name by clicking on the arrows corresponding to the month name. You can change the calendar into a monthly calendar, refer to Figure 3-4, by clicking on the month name.

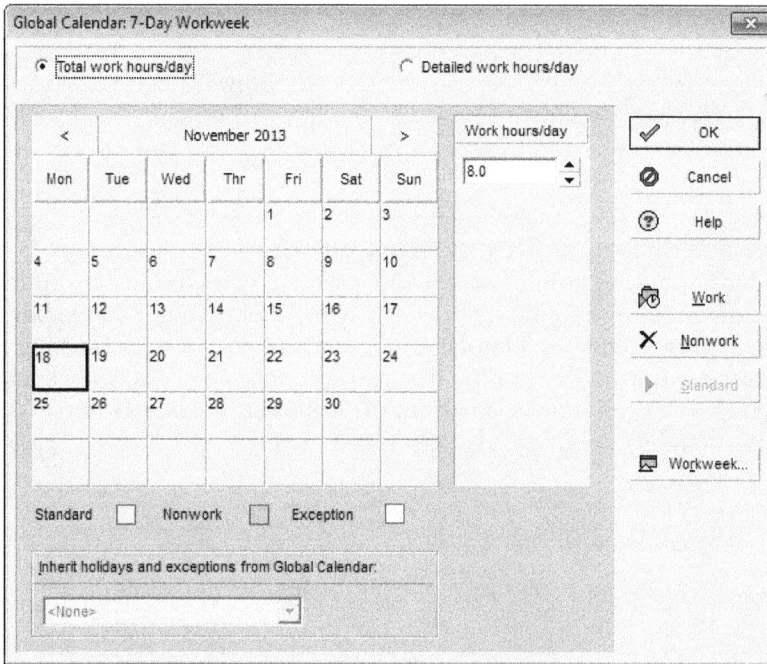

Figure 3-3   *The **Global Calendar: 7-Day Workweek** dialog box*

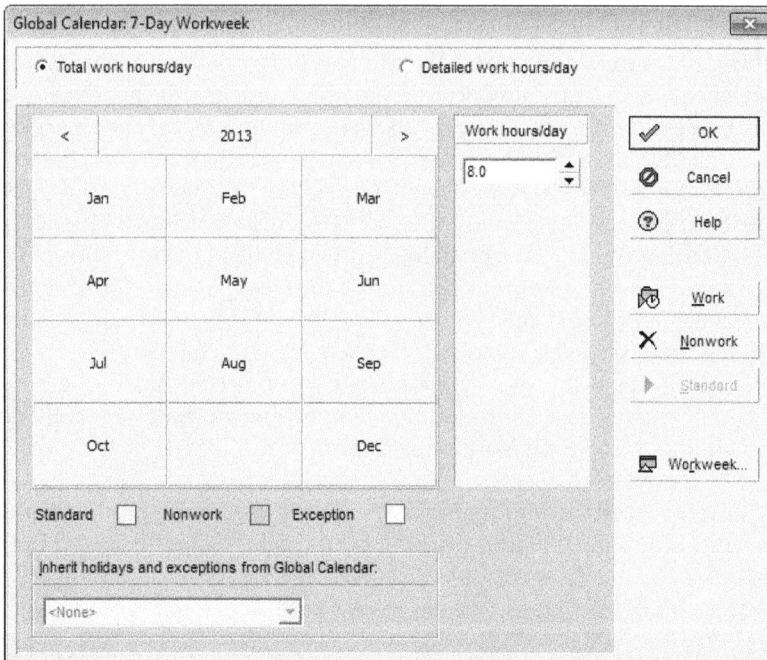

Figure 3-4   *The calendar displayed as a monthly calendar*

You can also select all the dates falling under a specific day of the month by clicking on the name of the day. For example, if you want to select all Saturdays in a month click on the **Sat**; all the saturdays of that month will be selected.

You can also change the workdays into non-working days. To do so, first highlight the day that you want to edit by clicking on the required day or highlight multiple days by using the **CTRL** key. After selecting the required day or days, choose the **Nonwork** button on the right in dialog box; the selected working day or days will be converted into non-working days. Similarly, by using the **Work** button, you can change a non-working day into a working day.

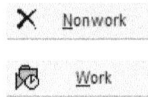

Modifying the Working Hours in the Calendar

The total number of working hours in a week are called as calendar weekly hours and they are calculated by adding the working hours in each day in a week. In the dialog box displayed for the selected calendar type and calendar name, you can adjust or edit the working hours of every workday of a calendar in two ways. In the first method, select the **Total work hours/day** radio button and then choose the **Workweek** button; the **Calendar Weekly Hours** dialog box will be displayed, as shown in Figure 3-5.

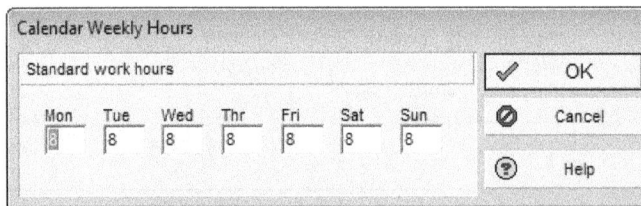

*Figure 3-5 The **Calendar Weekly Hours** dialog box displayed on selecting the **Total work hours/day** radio button*

In this dialog box, the number of work hours for each day of the week will be displayed. To edit the number of work hours for any desired day, click in the edit box corresponding to that day and specify the desired working hours. Next, choose the **OK** button to apply the changes.

In the second method, select the **Detailed work hours/day** radio button; the working hours for each work day will be displayed in detailed. Choose the **Workweek** button in the right side of the dialog box; the **Calendar Weekly Hours** dialog box will be displayed, as shown in Figure 3-6. In this dialog box, detailed hour sheet for each day of the week is available and each hour of the day is divided into two half hours slot. You can edit the workhours for any day by activating or deactivating the half hours cells. To activate or deactivate any cell, you have to double-click on the cell. For example, if you want to add 1 workhour to a day then select the time slot at which you need to add that extra 1 hour and double-click on the two halfhour cells; the specified 1 workhour will be added to the total workhour for that day.

Similarly, you can reduce the workhours for any required day. The cells for working hours will be displayed in white color and the non-working hours cells will be displayed in the same color as of the non-working days.

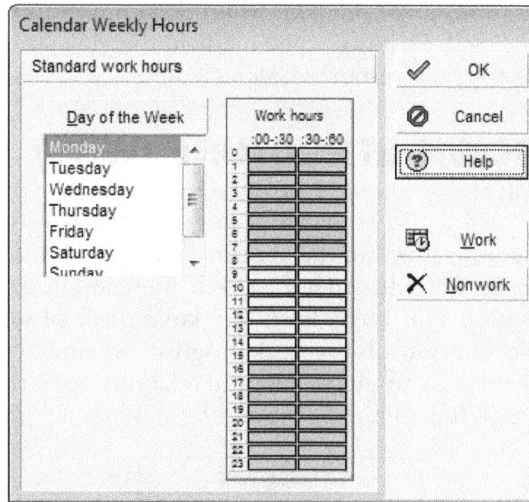

Figure 3-6 *The **Calendar Weekly Hours*** *dialog box displayed on selecting the* ***Detailed work hours/day*** *radio button*

Renaming and Deleting the Calendar

To rename a calendar, open the **Calendars** dialog box. In this dialog box, select the calendar type by selecting the required radio button (**Global/Resource/Project**) and then select the calendar to be renamed. Double-click on the selected calendar; the calendar name becomes editable. Specify the desired name in the edit box.

You can also delete a calendar if the calendar has no further use in the project. To do so, select the calendar type **Global/Resource/Project** in the **Calendars** dialog box and then select the required calendar to be deleted. After selecting the calendar, choose the **Delete** button from the right side of the dialog box; the **Primavera P6** message box will be displayed. Choose the **OK** button from the message box; the selected calendar will be deleted.

Note

1. If on a specific date, the number of hours entered does not match the number of workhours specified for that day, then the display color of the dates changes to white which indicates an exception in the calendar.

2. You cannot delete a calendar that is assigned to a project. If you try doing so, a message box will be displayed informing that this calendar is assigned to a project and cannot be deleted.

Inherit Holidays and Exceptions from a Global Calendar

When you are creating a new project or resource calendar, you also get an option to copy the calendar holidays of the global calendar to your new calendar. To do so, select the required calendar from the Global calendar list.

The newly created project or resource calendar will remain linked to the global calendar from which the holidays are inherited. So any change made in the global calendar holidays will also be reflected in the linked project or resource Calendar.

WORK BREAKDOWN STRUCTURE (WBS)

WBS in project management is a method of dividing a complex project into simpler and manageable tasks. In the real world scenario, a large project consists of many activities or levels and to complete the project within the specified cost and duration of time, you need to complete these levels piece by piece. In order to achieve these levels, the works are broken down into smaller tasks or activities. This breakdown of a large piece of work helps in simplifying the project work as a whole and controls the work progress. So, due to this breakdown of larger piece of work into smaller subtasks results in a logical relationships structure which is known as Work Breakdown Structure (WBS). Figure 3-7 shows the WBS for a typical construction project.

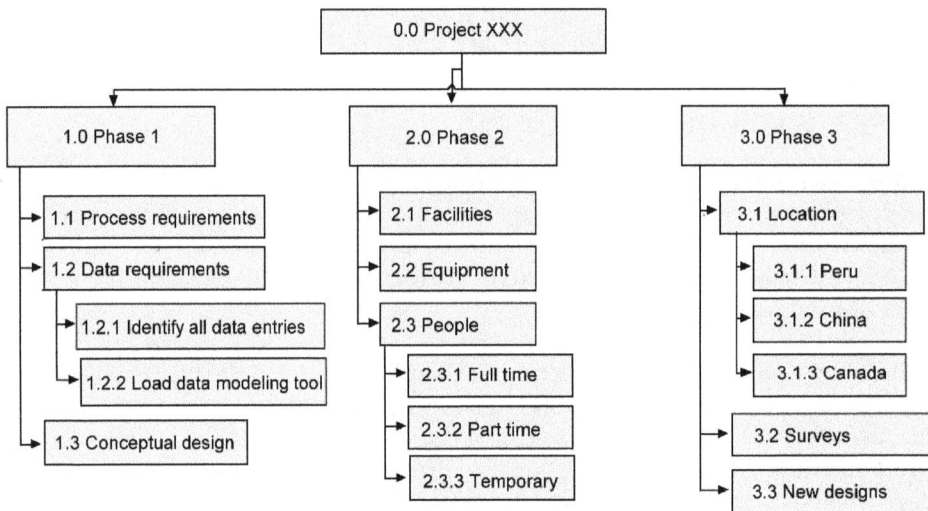

Figure 3-7 *The flowchart displaying the Work Breakdown Structure for a typical project*

So, Work Breakdown Structure is defined as the hierarchical arrangement of the activities to be followed during the manufacturing of a product. It can also represents the steps involved in a construction project.

Working with WBS in Primavera

In Primavera P6, the WBS enables you to organize a project into logical phases with the purpose of planning and control. It is a hierarchical structure in which activities are attached. A standard breakdown structure is pre-defined in Primavera P6 project templates, which helps in simplifying the concept of setting up the WBS as you do not have to create a new template for each new project.

In Primavera P6, the Work Breakdown Structure consists of levels representing the elements or the tasks needed to be accomplished for the completion of a project. The project is the root node for the Work Breakdown Structure and has a direct relationship with each WBS element. WBS is the summarized workflow of a project from its beginning to completion. You can view the whole WBS together, or can only view a specific node. Figure 3-8 shows the default process followed by Primavera P6 in creating WBS.

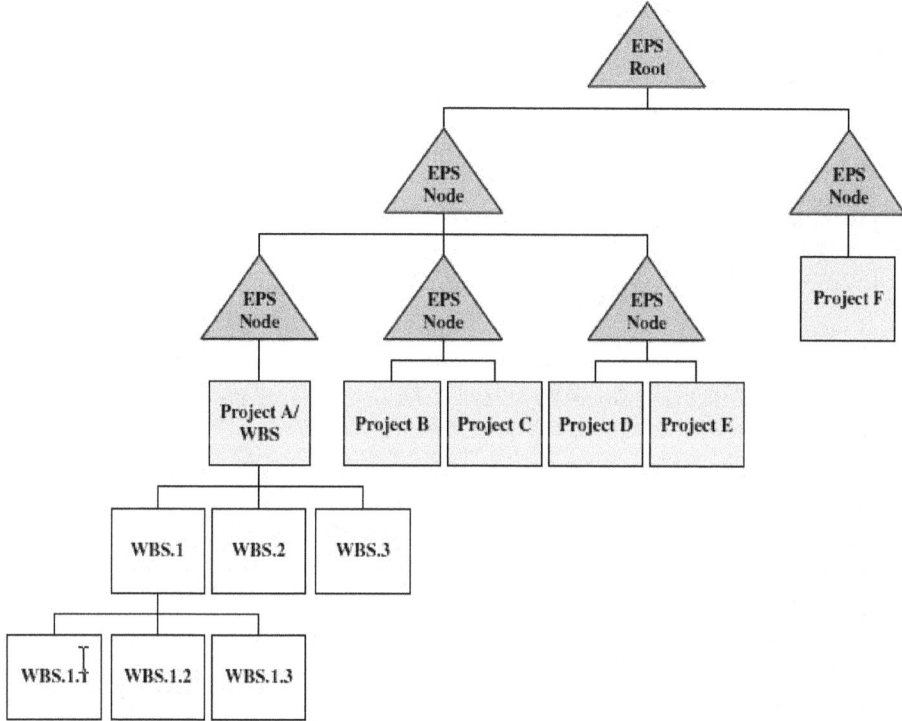

Figure 3-8 The default process followed by Primavera for creating WBS

Viewing the WBS for a Project

You can view the WBS for a project or an EPS node in the **Work Breakdown Structure** window. To display the window, open the required project and then choose the **WBS** option from the **Project** menu in the menubar or alternatively you can also select the **WBS** button from the Directory bar; the **Work Breakdown Structure** window for the selected project will be displayed, as shown in Figure 3-9.

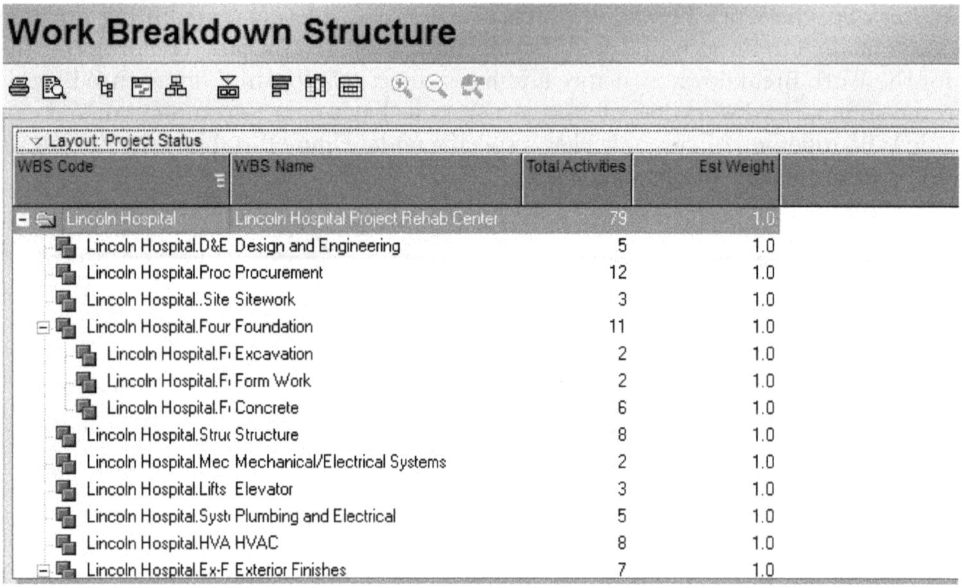
Work Breakdown Structure

Layout: Project Status

WBS Code	WBS Name	Total Activities	Est Weight
Lincoln Hospital	Lincoln Hospital Project Rehab Center	79	1.0
Lincoln Hospital.D&E	Design and Engineering	5	1.0
Lincoln Hospital.Proc	Procurement	12	1.0
Lincoln Hospital..Site	Sitework	3	1.0
Lincoln Hospital.Four	Foundation	11	1.0
Lincoln Hospital.F	Excavation	2	1.0
Lincoln Hospital.F	Form Work	2	1.0
Lincoln Hospital.F	Concrete	6	1.0
Lincoln Hospital.Struc	Structure	8	1.0
Lincoln Hospital.Mec	Mechanical/Electrical Systems	2	1.0
Lincoln Hospital.Lifts	Elevator	3	1.0
Lincoln Hospital.Syst	Plumbing and Electrical	5	1.0
Lincoln Hospital.HVA	HVAC	8	1.0
Lincoln Hospital.Ex-F	Exterior Finishes	7	1.0

Figure 3-9 *The* ***Work Breakdown Structure*** *window displaying the WBS for a project in a tabular form*

In Primavera P6, the Work Breakdown Structure for a project is divided into many sublevels known as WBS nodes, also known as WBS codes. Each WBS node is a single entity in a project structure to which activities are attached. Figure 3-10 shows the Work Breakdown Structure for a project consisting of different WBS nodes. It shows the Work Breakdown Structure for a project named as Edison High, which has WBS nodes such as Edison High 1, Edison High 2, and Edison High 3. WBS nodes are also further divided into WBS subnodes, for example Edison High WBS node has 4 WBS subnodes that are Edison High 1.1,1.2, 1.3, and 1.4. Each WBS node and subnode should have a unique and informative WBS name which will be displayed in the **WBS Code** column. In a WBS node, you can store various types of data such as anticipated dates, budget plan for the project, and monthly expenditure for the project. The financial information is also shared between projects and their WBS elements.

You can view the WBS for a project either as a chart or a table. Figure 3-9 shows the WBS in a tabular form. To view the WBS in the chart view, choose the **Chart View** button from the WBS toolbar in the **Work Breakdown Structure** window; the WBS will be displayed in a chart form. Figure 3-10 shows a WBS in the chart form.

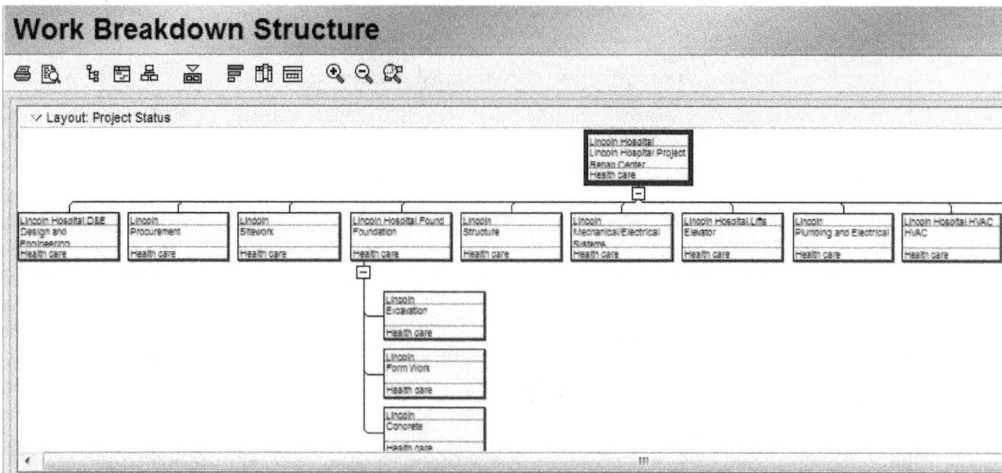

Figure 3-10 The **Work Breakdown Structure** *window displaying the WBS for a project in a chart form*

The buttons in the **WBS** toolbar help you in organizing the **Work Breakdown Structure** window. The various buttons in the WBS toolbar are discussed next.

WBS Table

You can choose the **WBS Table** button to display the WBS in a tabular form.

Chart View

You can choose the **Chart View** button to display the WBS in a chart form.

Gantt Chart

You can choose **Gantt Chart** button to view gantt chart in the **Work Breakdown Structure** window along with the WBS table or chart.

Show/Hide Bottom Layout

The **Show/Hide Bottom Layout** button is used to show or hide the **WBS Details** table displayed at the bottom of the **Project** window.

Bars

When you choose the **Bars** button from the WBS toolbar, the **Bars** dialog box will be displayed as shown in Figure 3-11. By using the options in the **Bars** dialog box, you can specify the style and labels for the bars in a gantt chart. The **Bars** dialog box contains three tabs: **Bar style**, **Bar Settings**, and **Bar Labels**.

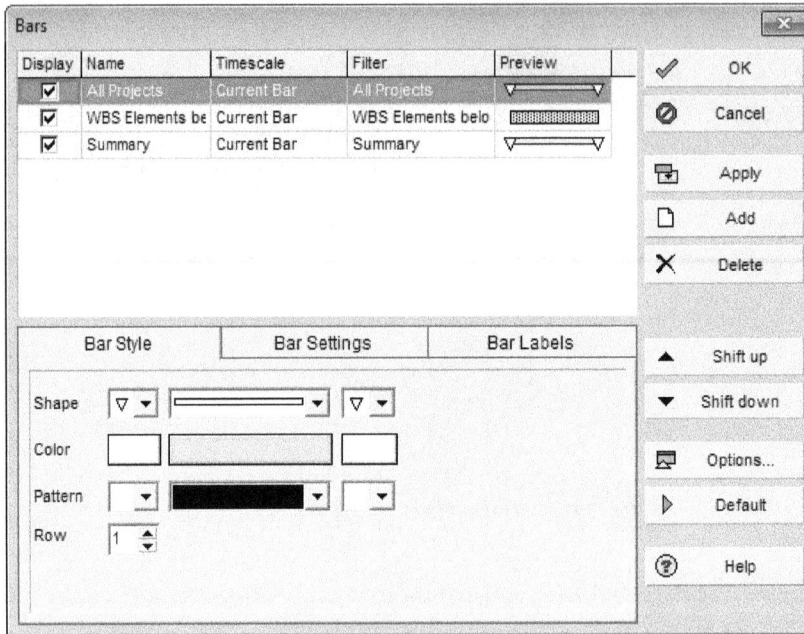

Figure 3-11 *The **Bars** dialog box with **Bar Style** tab chosen*

In the **Bar Style** tab, you can select the required option from the **Shape** drop-down lists to specify the shape of the start point and endpoints of the bar and also the required thickness of the bar. Similarly, you can specify color, pattern, and row for the bars in the Gantt chart by selecting the desired options.

In the **Bar Settings** tab, select the **Show bar when collapsed** check box to display the selected bar in the Gantt Chart even when the display is collapsed to the summary information level. Select the **Show bar for grouping bands** check box to display the selected bar as the summary bar.

In the **Bar Labels** tab, you can specify the desired position and database field of the selected bar's label. To do so, double-click on the required cell under the **Position** and **Label** columns; a drop-down list will be displayed. Choose the required option from the drop-down list. The standard structure for the position for a Gantt chart bar label is shown in Figure 3-12.

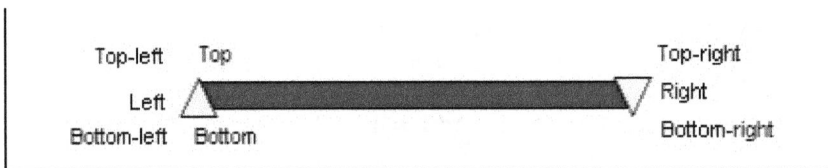

Figure 3-12 *Positions for a gantt chart bar label*

Columns

When you choose the **Columns** button from the WBS toolbar, the **Columns** dialog box will be displayed, as shown in Figure 3-13. In this dialog box, you can specify the columns that you want to display in your **Work Breakdown Structure** window. You can also specify the order in which the columns would be placed in the window.

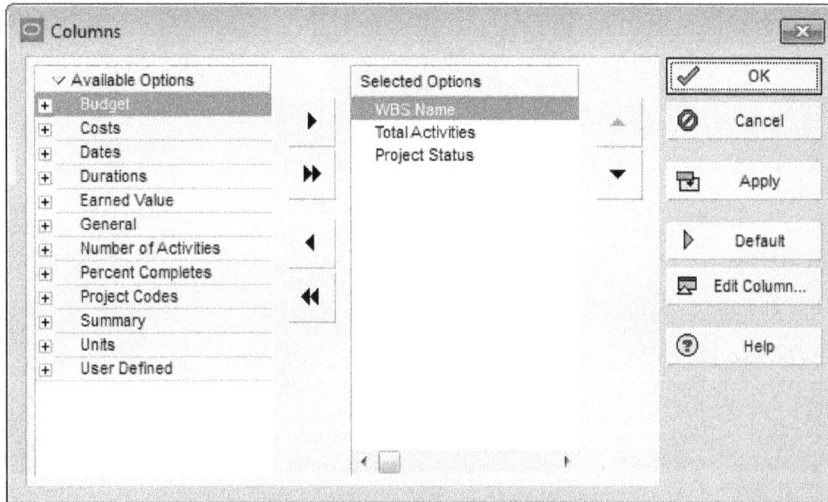

Figure 3-13 *The* **Columns** *dialog box*

To add or delete a column from your **Work Breakdown Structure** window, select the required option from the **Available Options** area or from the **Selected Options** area in the **Columns** dialog box and then choose the arrow keys in the dialog box to include or exclude the selected option. The **Selected Options** column displays the options which are currently displayed as columns in the **Work Breakdown Structure** window. Therefore when you add an option it will be reflected in the **Selected Options** column.

Timescale

When you choose the **Timescale** button, the **Timescale** dialog box will be displayed as shown in Figure 3-14. In this dialog box, you can specify the timescale you want to display in the current Gantt chart. In the **Timescale Format** area, you can set the number of time units to be displayed to either two or three by selecting the corresponding radio button. By using the options available in the **Date Format** area, you can specify the type, shift calendar, and date interval for the Gantt chart.

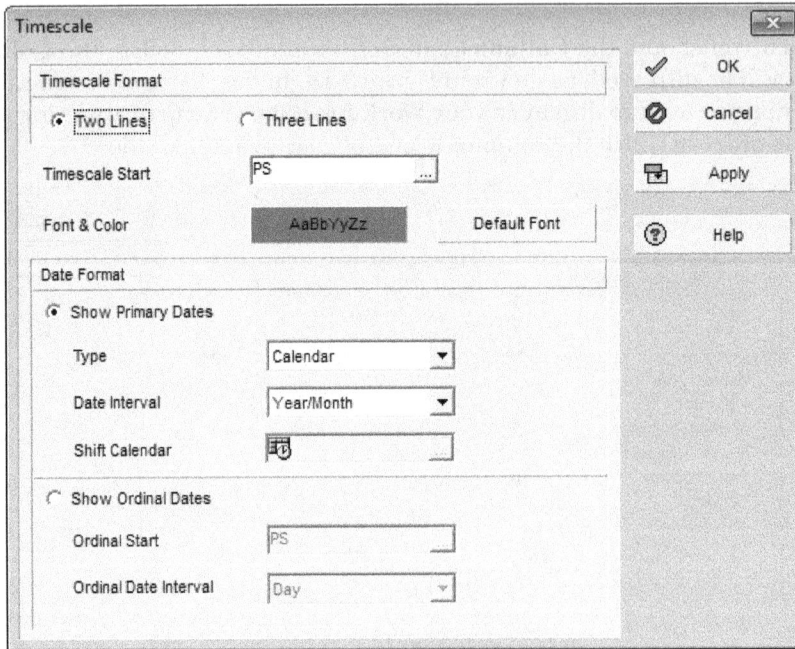

*Figure 3-14 The **Timescale** dialog box*

Creating a WBS and Adding WBS Elements

When a project is created, Primavera P6 automatically creates a root WBS element with the same name, EPS, and Project ID as that of the project. All the further WBS elements will be added under this root WBS element. Use the **Work Breakdown Structure** window to view all the WBS elements related to the required project. Also in this window, all the activities required to complete the project are added to the WBS elements. Typically on creating a project, the project manager first develops the WBS, assigns documents to each WBS element, and then defines the activities for performing the element's work. In addition to the activities and assignments, each WBS element is assigned a specific calendar and specific OBS element responsible for all the work included in the WBS element.

Adding a WBS Element

Before adding the WBS elements, you must be working in the **Activities** window that is organized based on all levels of WBS hierarchy, it is necessary that either you are working on a single project or working on a group of projects. Next, open the **Work Breakdown Structure** window by choosing the **WBS** button in the Directory bar. In this window, select the WBS element immediately above and under which you want to add the new element and then choose the **Add** button from the right side of the **Work Breakdown Structure** window; a new WBS element will be added and intended one level under the selected WBS element, refer to Figure 3-15.

You can also edit an existing WBS element by double clicking on the required WBS element and then rename it as desired. To outdent and indent an element's position in the WBS, use the arrow keys from the Command bar. To move the element upwards or downwards in the structure, use the up and down arrow keys.

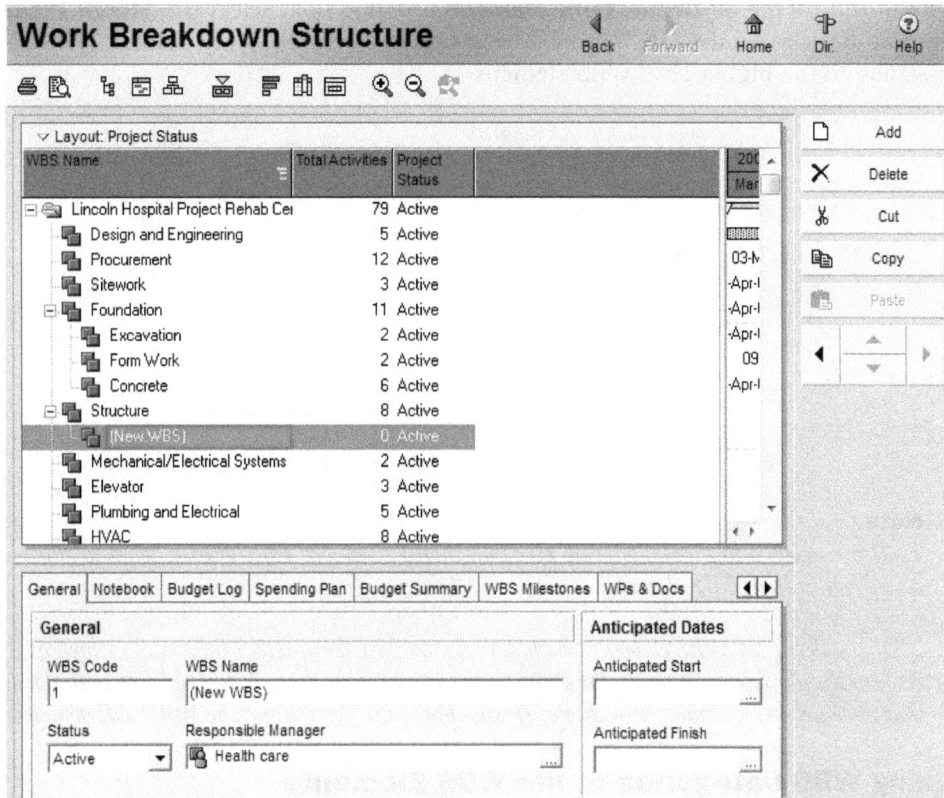

*Figure 3-15 The **Work Breakdown Structure** window displaying a new WBS element added*

Now, in the **General** tab of the **WBS Details** table displayed as the bottom layout, you can specify the code and name for the newly created element in the **WBS Code** and **WBS Name** edit box, respectively. In the **Status** drop-down, you can specify the status for the WBS element. To assign an OBS element to the newly created WBS element, choose the browse button next to the **Responsible Manager** edit box. On doing so, the **Select Responsible Manager** dialog box will be displayed. In this dialog box, select the OBS element to be assigned to your WBS element and choose the **Select** button from the right in the dialog box; the selected OBS element will be assigned.

In the **Notebook** tab, you can add and save notes related to the WBS element. In the **Budget Log** tab, you can specify the settings related to your specified budget and can also view the current and the proposed budget for the project. In the **Budget Summary** tab, you can view the summary of your project budget. All the areas of this tab are calculated from the values entered in the **Budget Log** tab.

Deleting a WBS Element
You can also delete a WBS element. To do so, select the WBS element you want to delete and then choose the **Delete** button from the right side of the **Work Breakdown Structure** window; the **Merge or Delete WBS Element(s)** dialog box will be displayed, as shown in Figure 3-16. If the WBS element has activity assignments, then you also have the option to merge the activity assignments of the WBS element to be deleted with the element's higher level WBS element.

To do so, in the **Merge or Delete WBS Element(s)** dialog box, select the **Merge Element(s)** radio button and choose the **OK** button. The assigned activities of the deleted WBS element are reassigned to the higher level WBS element.

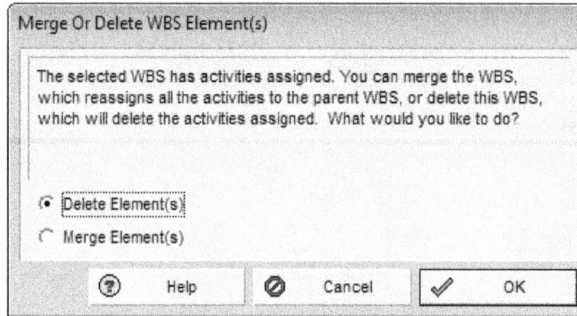

Figure 3-16   The Merge Or Delete WBS Element(s) dialog box

Note

1. If you delete a higher level WBS element, then Primavera P6 will also delete all the elements contained in that WBS element.

*2. If you delete the WBS element which does not contain any further information such as activity, resources and role assignments, the **Primavera P6** message box will be displayed with the message that once deleted it cannot be undone. If you choose the **Yes** button, the WBS will be deleted.*

Assigning WBS Categories to the WBS Elements

The WBS categories for the WBS elements are same as the activity codes for activities. WBS categories help the WBS elements of a project to be grouped and sorted. You can create WBS categories by using the **Admin Categories** dialog box. To invoke this dialog box, choose the **Admin Categories** option from the **Admin** menu in the **Menu** bar. In the **Admin Categories** dialog box, choose the **Project Phase** tab; the **WBS Categories** page will be displayed, as shown in the Figure 3-17. In this page, several WBS categories will be displayed and also you can create a new WBS category. After creating and editing, you can assign these values to WBS elements.

Note

*In Primavera P6, the **Project Phase** is the default WBS category value.*

WBS category values can be assigned to the WBS elements in the **Project Phase** column of the **Work Breakdown Structure** window. To show/hide the **Project Phase** column, you can use the **Columns** dialog box. To invoke the dialog box, choose the **Columns** button from the WBS toolbar; the **Columns** dialog box will be displayed, refer to Figure 3-13. In this dialog box, add the desired column to the **Work Breakdown Structure** window. Now, select the WBS element to which you want to assign a category and, then choose the Browse button in the **Project Phase** column.

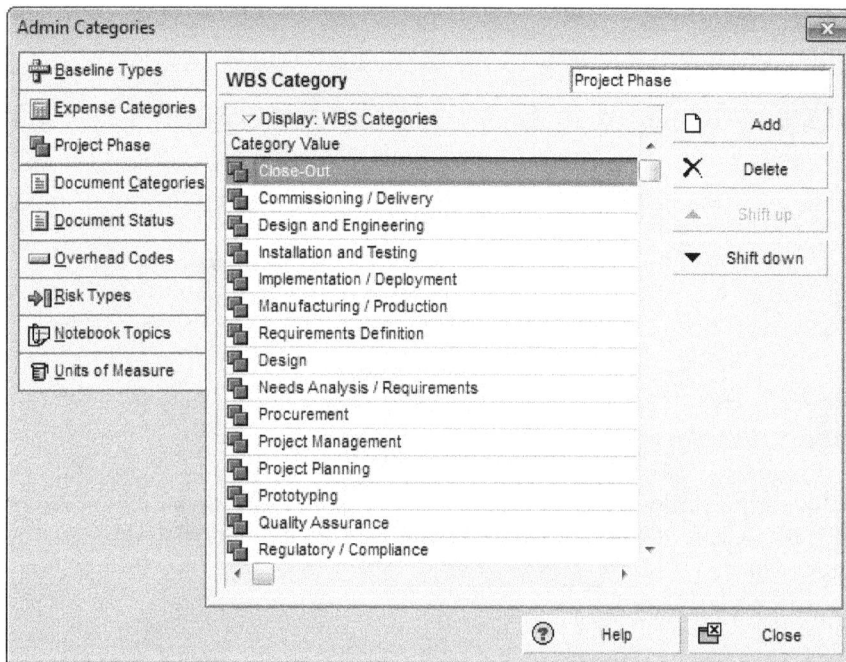

Figure 3-17 *The **WBS Category** page displayed in the **Admin Categories** dialog box*

Defining Earned Value Settings for WBS Elements

Earned value is a technique for monitoring the project performance according to both the project costs and schedule. It informs you about what you have actually earned against what you spent. This technique compares the budgeted cost of the work to the actual cost. The general formula for computing the earned value is given below:

Earned Value (EV) = Budget at Completion (BAC) x Performance % Complete

The earned value analyses is typically performed for WBS elements and to perform the analysis you first need to define earned value settings for the WBS element. To do so, select the WBS element whose earned value settings you want to define and then choose the **Earned Value** tab from the **WBS Details** table; the options under the **Earned Value** tab will be displayed, as shown in Figure 3-18. The options under this tab have been discussed next.

Technique for computing performance percent complete

The options in the **Technique for computing performance percent complete** area are used to specify the technique required for calculating percent complete. Some of the techniques for calculating percent complete are discussed next.

Activity percent complete

This technique is used to calculate earned value according to the current activity percentage complete. If you select the **Use resource curves / future period buckets** check box, then the activity percent complete technique will override all the activities that have a resource curve assigned to at least one of the resource assignments, or the activities that have assignments with manually defined future period bucket values.

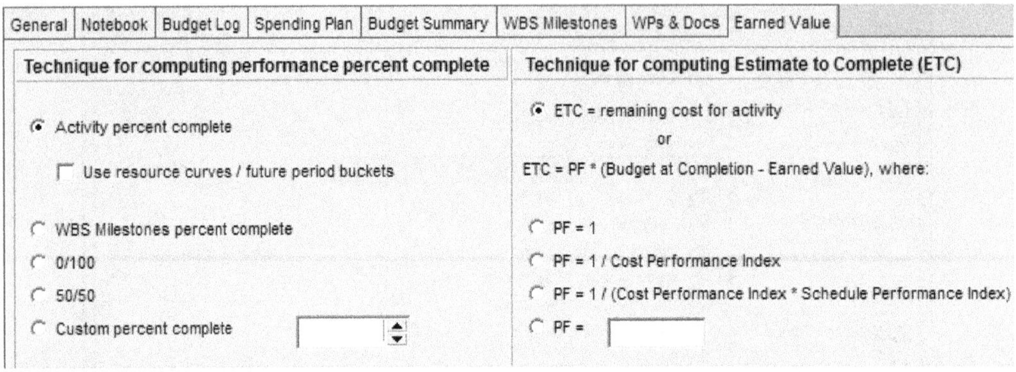

| General | Notebook | Budget Log | Spending Plan | Budget Summary | WBS Milestones | WPs & Docs | Earned Value | |

Technique for computing performance percent complete   **Technique for computing Estimate to Complete (ETC)**

     (•) ETC = remaining cost for activity

(•) Activity percent complete   or

    ☐ Use resource curves / future period buckets   ETC = PF * (Budget at Completion - Earned Value), where:

( ) WBS Milestones percent complete   ( ) PF = 1

( ) 0/100   ( ) PF = 1 / Cost Performance Index

( ) 50/50   ( ) PF = 1 / (Cost Performance Index * Schedule Performance Index)

( ) Custom percent complete   [▲▼]   ( ) PF = [   ]

Figure 3-18 *The various options in the* ***Earned Value*** *tab used for earned value analysis*

WBS Milestones percent complete

This technique is used to calculate earned value according to completion of the WBS element's weighted milestones. As the project progresses, you mark each milestone complete and the WBS's element's performance percent complete is calculated based on the weight of the milestone.

0/100

This technique will calculate earned value only when the activities under the selected WBS element are 100% complete.

50/50

This technique starts calculating the earned value when the activity is 50% complete and will calculate until the activity is complete.

Technique for computing Estimate to Complete (ETC)

The options in the **Technique for computing Estimate to Complete (ETC)** area are used to specify the technique required which will be used on calculating an activity's estimate to complete the (ETC) value. Some of the techniques for calculating ETC are discussed next.

ETC = remaining cost for activity

This technique is chosen by default for calculating ETC. ETC is used as the remaining cost of the activity. This technique does not affect EV in calculating your ETC.

ETC = PF * (Budget at Completion - Earned Value)

This technique calculates ETC by multiplying the Performance factor (PF) with the value obtained by subtracting the Earned value from the value of the total budget at the completion of the activity. The Performance factor is a user- selected value. The formulas to calculate the Performance factor value given in the **Technique for computing Estimate to Complete (ETC)** area are listed below:

PF=1

PF=1/Cost Performance Index

PF= 1/ (Cost Performance Index *Schedule Performance Index)

In the above listed formulas, the **Cost Performance Index (CPI)** is the earned value cost divided by the actual cost and the **Scheduled Performance Index (SPI)** is the earned value cost divided by the planned value cost.

PF value can also be directly specified by you. To do so, specify the required value in the **PF** edit box. Choose any of the desired method by selecting the radio button for that method.

WBS Milestones

In the early stages of project planning, the person or the team that is responsible for project planning and for the project processes need to decide how Primavera P6 will calculate earned value, percent complete, resource use and financial data.

In Primavera P6, you can calculate the performance percent complete for an activity by using WBS milestones which are further used for computing the earned value. You can add unlimited number of milestones to a WBS element and each milestone is given a weight that indicates its importance to the project progress. When a WBS milestone is marked as complete, Primavera uses the weight assigned to that milestone for calculating the performance percent complete for all the activities under that WBS element.

For example, if a WBS element has 4 activities under it and 4 WBS milestones are assigned to that WBS element, then completion of each milestone will add 25% to the performance percentage complete for all the activities.

Adding WBS Milestones

To add WBS milestones to a WBS element, choose the **WBS Milestones** tab from the **WBS details** table in the **Work Breakdown Structure** window; the various options of this tab will be displayed, as shown in the Figure 3-19.

Using these options you can add unlimited number of milestones to your project. Choose the **Add** button to add a WBS milestone to your project. To delete a WBS milestone, select the milestone to be deleted and then choose the **Delete** button corresponding to the **Add** button; the **Primavera P6** message box is displayed prompting that are you sure about deleting the WBS milestone. If you choose the **Yes** button; the selected milestone will be deleted.

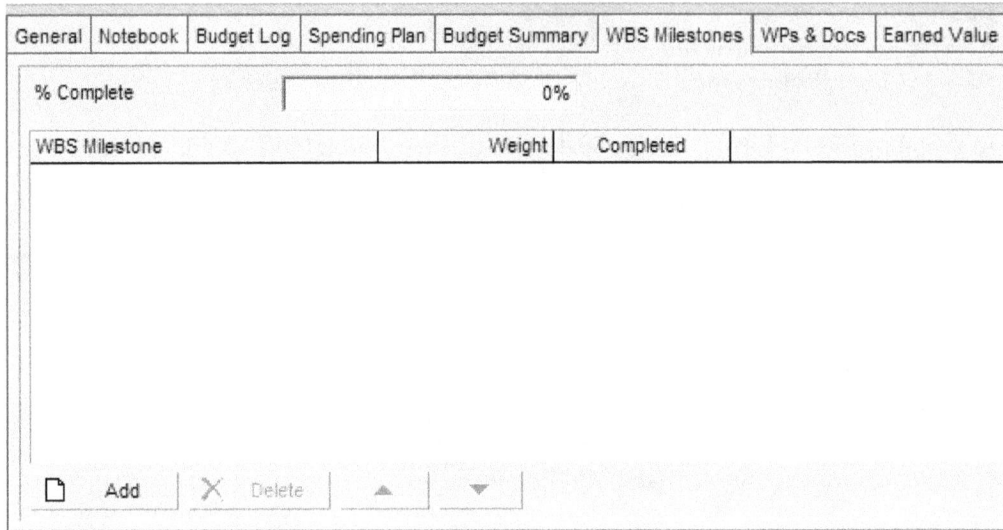

Figure 3-19 *The various options in the **WBS Milestones** tab*

TUTORIALS

To perform tutorials of this chapter, you need to complete Tutorial 1 and Tutorial 2 of Chapter 2.

If you have not performed Tutorial 2 of Chapter 2, then you need to download the tutorial file from *www.cadcim.com* by using the following steps:

1. Browse to *Textbooks > Civil/GIS> Primavera P6 > Exploring Oracle Primavera P6 v7.0*. Next, select the *c02_Primavera_P6_tut.zip* file from the **Tutorial Files** drop-down list. Choose the corresponding **Download** button to download the data file.

2. Now, save and extract the downloaded folder to the following location:

    C:/PM6

3. Import the downloaded files under the EPS which is created in Tutorial 1 of Chapter 2.

Tutorial 1	Home Project WBS

In this tutorial, you will create WBS for the **Home Construction** project created in Tutorial 2 of Chapter 2.   **(Expected time: 30 min)**

The following steps are required to complete this tutorial:

a. Open the **Projects** window and the project.
b. Create work breakdown structure for the project.
c. Export the project.

Opening the Projects Window and the Project

1. Start Primavera P6 and invoke the **Projects** window by choosing the **Projects** option from the **Enterprise** menu in the menubar.

2. In this window, select the **Home Construction** project placed under the **Construction** EPS node and right-click; a shortcut menu is displayed.

3. Choose the **Open Project** option from this menu; the **Activities** window with the project **Home Construction** is displayed.

Creating the Work Breakdown Structure Window

1. In the **Activities** window, choose the **WBS** button from the Directory bar; the **Work Breakdown Structure** window for the **Home Construction** project is displayed.

2. Choose the **Columns** button from the WBS toolbar; the **Columns** dialog box is displayed.

3. Select the **Node Type, Project Phase, Project Status**, and **Responsible Manager** option under the **General** node from the **Available Options** area and move them to the **Selected Options** area and then choose the **OK** button.

4. In the **Selected Options** area, arrange the options, as shown in Figure 3-20.

Note
The options should be arranged, as shown in the Figure 3-20. Other options should be moved from the Selected Options area to the Available Options area.

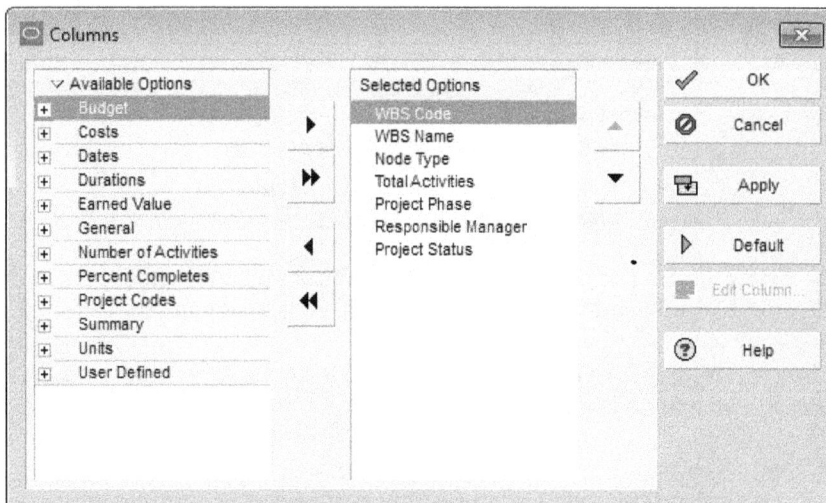

*Figure 3-20 The **Columns** dialog box with the arranged options*

5. Next, double-click on the field corresponding to the **Home Construction** project under the **Project Phase** column; the **Select Project Phase** dialog box is displayed.

6. In this dialog box, select the **Design and Engineering** option from the **Category Value** list and choose the **Select** button from the right side in the dialog box; the **Select Project Phase** dialog box is closed and the phase is set to design & engineering.

7. Choose the **Add** button from the Command bar; a new WBS node with name **New WBS** is added under the **Home Construction** project node.

8. Specify **Grading** as the name of the created WBS; the WBS Code is entered according to the project.

9. Double-click under the **Responsible Manager** column; the **Select Responsible Manager** dialog box is displayed, as shown in Figure 3-21.

10. Ensure that the **All OBS Elements** option is displayed in the **Display** drop-down list, refer to Figure 3-21. As a result, the **All OBS Elements** list is displayed in the **OBS Name** area.

Figure 3-21 The Select Responsible Manager dialog box with the All OBS Elements option selected

11. Select the **Site Engineer** option from the **OBS Name** area and choose the **Select** button; the dialog box is closed and the responsible manager is assigned.

12. Repeat the procedure followed in steps 4 through 8 to create other WBS elements and assign the responsible manager, as shown in Figure 3-22.

Work Breakdown Structure

WBS Code	WBS Name	Node Type	Total Activities	Project Phase	Responsible Manager	Project Status
⊟ CD00720	Home Construction	Project	0	Design and Engineering	Project Manager	Active
CD00720.1	Grading	WBS	0	Design and Engineering	Site Engineer	Active
CD00720.2	Foundation	WBS	0	Design and Engineering	Site Engineer	Active
⊟ CD00720.3	Structure	WBS	0	Design and Engineering	Site Engineer	Active
CD00720.3.1	Framing	WBS	0	Design and Engineering	Civil Foreman	Active
CD00720.3.2	Roof	WBS	0	Design and Engineering	Civil Foreman	Active
⊟ CD00720.3.3	Interior	WBS	0	Design and Engineering	Site Engineer	Active
CD00720.3.3.1	Mechanical	WBS	0	Design and Engineering	Civil Foreman	Active
CD00720.3.3.2	Electrical	WBS	0	Design and Engineering	Civil Foreman	Active
CD00720.3.3.3	Plumbing	WBS	0	Design and Engineering	Civil Foreman	Active
CD00720.3.3.4	Decor	WBS	0	Design and Engineering	Site Engineer	Active
CD00720.3.4	Exterior	WBS	0	Design and Engineering	Site Engineer	Active
CD00720.3.5	Landscaping	WBS	0	Design and Engineering	Site Engineer	Active

*Figure 3-22 The WBS elements added under the **Home Construction** project node*

13. With the help of the arrow keys, arrange the order of the WBS elements, refer to Figure 3-21.

    The WBS is created and is assigned to the project. Now, the project is saved in primavera and needs to be exported as an .XER files.

Exporting Projects

Export projects to save the project file for future reference.

1. To export the **Home Construction** project, ensure that the **Home Construction** project is opened in Primavera P6. If not, then select the **Home Construction** project from the **Projects** window and right-click; a menu is displayed. Choose the **Open Project** option from the menu.

2. Now, choose the **Export** option from the **File** menu; the **Export** wizard with the **Export Format** page is displayed.

3. In this page, ensure that the **Primavera PM/MM - (XER)** radio button is selected and then choose the **Next** button; the **Export Type** page is displayed.

4. In this page, ensure that the **Project** radio button is selected and then choose the **Next** button; the **Projects To Export** page is displayed.

5. Make sure the **Home Construction** project is displayed in this page. Choose the **Next** button; the **File Name** page is displayed.

6. In this page, choose the Browse button adjacent to the **File Name** edit box; the **Save File** dialog box is displayed.

7. In this dialog box, browse to the *C:\PM6* folder and then create a sub-folder with the name **c03**. Next, open the created folder and save the file with the name **c03_CONS_Home_tut01** and then choose the **Save** button.

8. Choose the **Finish** button; the **Primavera P6** message box is displayed containing the message that the export was successful.

9. Choose the **OK** button; the file is exported.

Tutorial 2   Calendars

In this tutorial, you will create calendars with the constraints listed below:

Calendar 1
      Working hours- 9:00-18:00 (9 workhours/day)
      Weekly off- Sunday and Monday.
Calendar 2
      Working Hours- 24 x 7 hours shift.
      Weekly Off-   No off
Also, you have to assign Calendar1 to the specified project.   **(Expected time: 30 min)**

The following steps are required to complete this tutorial:

a. Open the **Projects** window.
b. Create Calendar 1 and Calendar 2.
c. Assign Calendar 1 to the project.
d. Export the project.

Opening the Projects Window

1. Start Primavera P6 and invoke the **Projects** window by choosing the **Project** option from the **Enterprise** menu in the menubar.

2. In this window, choose the **Calendars** option from the **Enterprise** menu in the menubar, refer to Figure 3-23; the **Calendars** dialog box is displayed.

Creating the Calendars

In the first part of this section, you will create **Calendar 1**.

1. In the **Calendars** dialog box, select the **Project** radio button; the **Display: Project Calendars** area is displayed.

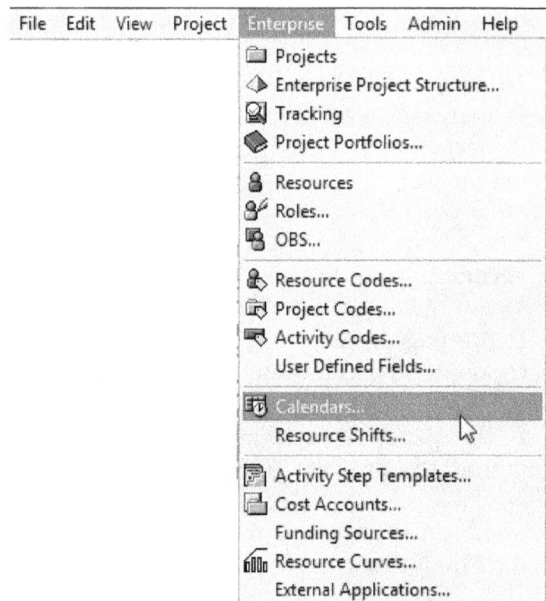

Figure 3-23 *Choosing the* **Calendars** *option from the* **Enterprise** *menu*

2. Choose the **Add** button from the right side of the dialog box; the **Select Calendar To Copy From** dialog box is displayed.

3. In this dialog box, select the **Standard 5 Day Workweek** calendar from the **Calendar Name** column and choose the **Select** button from the right side of the dialog box.

4. The **Select Calendar To Copy From** dialog box is closed and a new calendar is added under the **Display: Project Calendars** area of the **Calendars** dialog box.

5. Click on the calendar name; the name box becomes editable. Name this calendar as **Calendar 1**.

6. Now, choose the **Modify** button from the right side of the dialog box; the **Project Calendar: Calendar 1** dialog box is displayed, as shown in Figure 3-24.

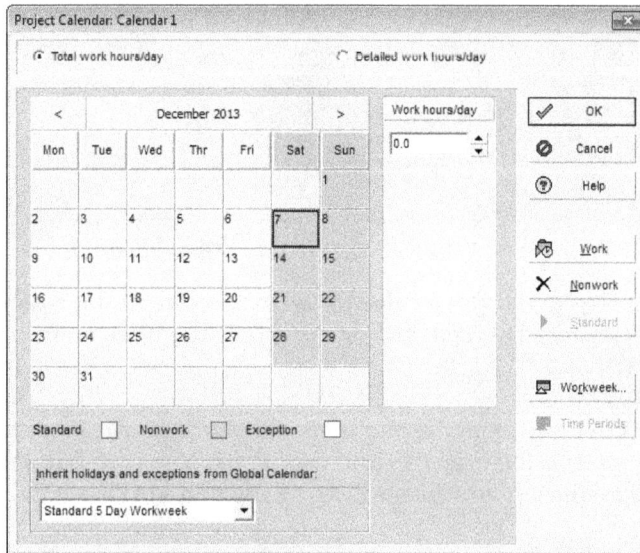

*Figure 3-24 The **Project Calendar: Calendar1** dialog box*

7. Choose the **Workweek** button from the right side of this dialog box; the **Calendar Weekly Hours** dialog box is displayed, as shown in Figure 3-25.

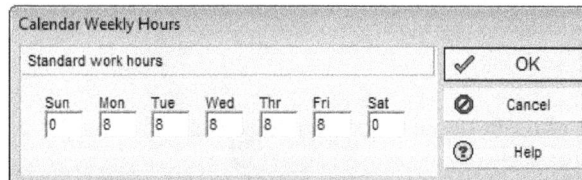

*Figure 3-25 The **Calendar Weekly Hours** dialog box*

8. In this dialog box, enter **9** in the **Tue** to **Sat** edit boxes and **0** in the **Mon** and **Sun** edit boxes as the standard working hours.

9. Choose the **OK** button; the **Calendar Weekly Hours** dialog box is closed and Saturday is standardized as a working day and Monday as a non-working day.

10. In the **Project Calendar: Calendar 1** dialog box, select the **Detailed work hours/day** radio button and then choose the **Workweek** button from the right side of the dialog box; the **Calendar Weekly Hours** dialog box is displayed, as shown in Figure 3-26.

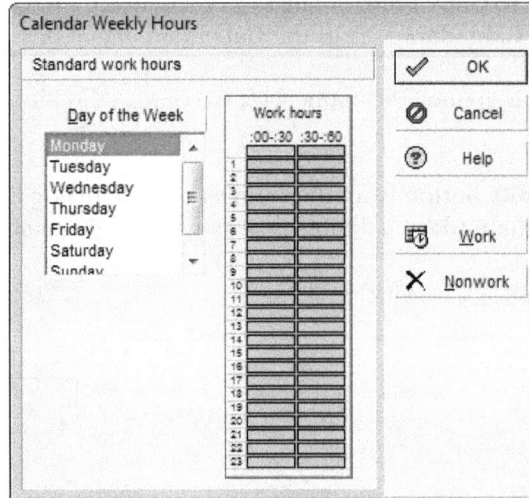

*Figure 3-26 The **Calendar Weekly Hours** dialog box*

11. In this dialog box, from the **Day of the Week** area, select all the days except Sunday and Monday and then select all the cells between 9 to 18 in the **Work hours** area using the CTRL key.

12. Choose the **Work** button; all the selected cells turns white and are converted into working hours. Similarly, select cell number 14 and choose the **Nonwork** button; the cell turns grey and is converted as lunch hour which will act as the non-working hour.

13. Select cell numbers 7 and 8 and then choose the **Nonwork** button; these cells are converted as non working hours.

14. Choose the **OK** button from the right side of the dialog box; the **Calendar Weekly Hours** dialog box is closed and 9:00-18:00 hours is assigned as the work shift on the working days.

15. Choose the **OK** button from the right side of the **Project Calendar: Calendar 1** dialog box; the dialog box is closed.

    In this second part of the tutorial, you will create **Calendar 2**.

16. Choose the **Add** button from the right side of the **Calendars** dialog box; the **Select Calendar To Copy From** dialog box is displayed.

17. Select the **7x24 hr. Days** calendar from the **Calendar Name** column and choose the **Select** button from the right side of the dialog box; a new calendar is added under the **Project Calendars** area of the **Calendars** dialog box. Name this calendar as **Calendar 2**.

18. Choose the **Modify** button; the **Project Calendar: Calendar 2** dialog box is displayed. Select the **Total work hours/day** radio button.

19. Choose the **Workweek** button from the right side of this dialog box; the **Calendar Weekly Hours** dialog box is displayed.

20. Ensure that 24 is entered in the edit boxes of all the Days (Mon through Sun) and choose the **OK** button; the dialog box is closed.

21. Select the **Detailed work hours/day** radio button and then choose the **Workweek** button; the **Calendar Weekly Hours** dialog box is displayed.

22. In this dialog box, choose the days from Sunday to Saturday using the SHIFT key from the **Day of the Week** area.

23. Select the cell number 4, 12, 20 using the **CTRL** key and then choose the **Nonwork** button.

24. Choose the **OK** button twice; the dialog box is closed and the **Calendars** dialog box is displayed with the modified calendars.

25. Select **Calendar 1** from the **Project Calendars** area and then choose the **To Global** button; the **Primavera P6** message box is displayed informing you to convert the project calendar into global.

26. Choose the **Yes** button; the project calendars are converted into global calendars.

27. Similarly, convert other project calendar, **Calendar 2** to global calendar.

28. Now, choose the **Close** button; the **Calendars** dialog box is closed.

Assigning the Calendar 1 to the Project

1. Choose the **Projects** option from the **Enterprise** menu in the menubar; the **Projects** window is displayed.

2. In this window, select the **Home Construction** project and then choose the **Defaults** tab from the **Project Details** table.

> **Note**
> *If the Project Details table is not displayed at the bottom of the Projects window, then right-click in the window; a shortcut menu is displayed, as shown in Figure 3-27. Choose the Project Details option from this menu; the Project Details table is displayed at the bottom of the Projects window.*

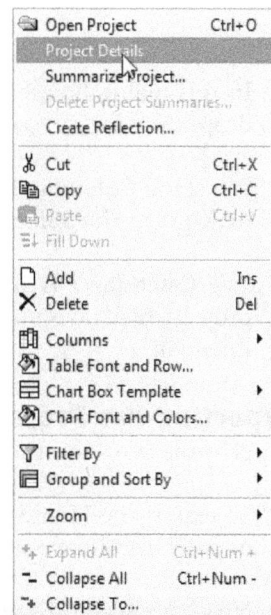

Figure 3-27 Choosing the Project Details option from the shortcut menu

Note

*If the **Defaults** tab is not displayed in the **Project Details** table then right-click anywhere in the details table; a menu is displayed. Choose the **Customize Project Details** option from the menu; the **Project Details** dialog box is displayed. In this dialog box, choose the **Defaults** option from the Available Tabs column and then shift it under the **Display Tabs** column and choose the **OK** button.*

3. In this tab, choose the Browse button adjacent to the **Calendar** edit box; the **Select Default Project Calendar** dialog box is displayed, as shown in Figure 3-28.

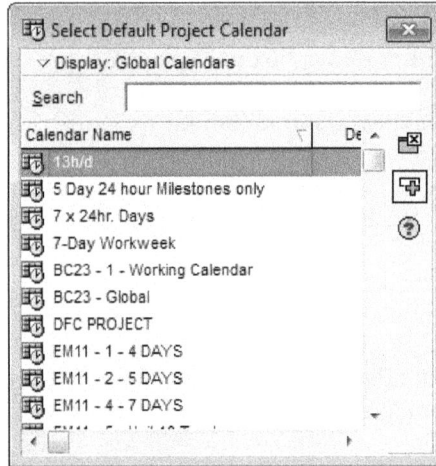

***Figure 3-28** The Select Default Project Calendar dialog box*

4. In this dialog box, ensure that the **Global Calendars** option is displayed in the **Display** drop-down list.

5. Select the **Calendar 1** option and then choose the **Select** button from the right side of the dialog box; the **Select Default Project Calendar** dialog box is closed.

    The **Calendar 1** is assigned to the project **New Home Construction**. Any activity in this project which does not have an assigned calendar will have **Calendar 1** assigned as its default calendar.

Exporting the Projects

Export project to save the project file for future reference.

1. To export the **Home Construction** project, ensure that the **Home Construction** project is opened in Primavera P6.

2. Now, choose the **Export** option from the **File** menu; the **Export** wizard with the **Export Format** page is displayed.

3. In this page, ensure that the **Primavera PM/MM - (XER)** radio button is selected and then choose the **Next** button; the **Export Type** page is displayed.

4. In this page, ensure that the **Project** radio button is selected and then choose the **Next** button; the **Projects To Export** page is displayed.

5. Make sure the **Home Construction** project is displayed in this page. Choose the **Next** button; the **File Name** page is displayed.

6. In this page, choose the Browse button adjacent to the **File Name** edit box; the **Save File** dialog box is displayed.

7. In this dialog box, browse to the *C:\PM6\c03* folder and save the file with the name **c03_CONS_Home_tut02** and then choose the **Save** button.

8. Choose the **Finish** button; the **Primavera P6** message box is displayed with the message that the export was successful.

9. Choose the **OK** button; the file is exported.

Tutorial 3   Banquet Hall WBS

In this tutorial, you will create WBS of highway project and then assign the calendar.

(Expected time: 30min)

The following steps are required to complete this tutorial:

a. Open the **Projects** window.
b. Create Work Breakdown Structure.
c. Assign **Calendar 2** to the project.
d. Export the project.

Opening the Projects Window

1. Start Primavera P6 and invoke the **Projects** window by choosing the **Project** option from the **Enterprise** menu in the menubar.

2. In this window, select the **Banquet Hall Construction** project placed under the **Consultancy Division** EPS node and right-click; a shortcut menu is displayed.

3. Choose the **Open Project** option from this menu; the **Activities** window with the project **Banquet Hall Construction** project is displayed.

Creating the Work Breakdown Structure

1. In the **Activities** window, choose the **WBS** button from the Directory bar; the **Work Breakdown Structure** window for the **Banquet Hall Construction** project is displayed.

2. Select the **Banquet Hall Construction** project node and double-click under the **Responsible Manager** column; the **Select Responsible Manager** dialog box is displayed.

3. Ensure that the **All OBS Elements** option is displayed in the **Display** drop-down list. Select the **Consultancy Head** as the responsible manager from the list and then choose the **Select** button; the dialog box is closed and the manager is assigned.

4. Double-click in the edit box under the **Project Phase** column of the **Work Breakdown Structure** window; the **Select Project Phase** dialog box is displayed.

5. In this dialog box, select the **Design and Engineering** option and choose the **Select** button from the right side; the **Select Project Phase** dialog box is closed and the phase is set to design and engineering.

6. Choose the **Add** button from the Command bar; a new WBS node with name **New WBS** is added under the **Banquet Hall Construction** project node.

7. Specify **Earthwork** as the name of the created WBS; Primavera P6 generates the WBS code according to the code given to the project.

8. Double-click under the **Responsible Manager** column; the **Select Responsible Manager** dialog box is displayed.

9. In the **OBS Name** list, select the **Engineers** option and choose the **Select** button; the dialog box is closed and the responsible manager is assigned.

10. Repeat the procedure followed in steps 4 through 9 to create other WBS elements and to assign the responsible manager, refer to Figure 3-29.

11. With the help of the arrow keys arrange the order of the WBS elements, refer to Figure 3-29.

Work Breakdown Structure

Layout: WBS

WBS Code	WBS Name	Node Type	Total Activities	Project Phase	Responsible Manager	Project Status
CN00770	Banquet Hall Construction	Project	0	Design and Engineering	Consultancy Head	Active
CN00770.1	Earthwork	WBS	0	Design and Engineering	Engineers	Active
CN00770.2	Substructure	WBS	0	Design and Engineering	Engineers	Active
CN00770.2.1	Concreting	WBS	0	Design and Engineering	Engineers	Active
CN00770.2.2	Shuttering	WBS	0	Design and Engineering	Engineers	Active
CN00770.2.3	Reinforcement	WBS	0	Design and Engineering	Engineers	Active
CN00770.3	Roadwork	WBS	0	Design and Engineering	Engineers	Active
CN00770.4	Superstructure	WBS	0	Design and Engineering	Engineers	Active
CN00770.4.1	Reinforcement	WBS	0	Design and Engineering	Engineers	Active
CN00770.4.2	Concreting	WBS	0	Design and Engineering	Engineers	Active
CN00770.4.3	Shuttering	WBS	0	Design and Engineering	Engineers	Active
CN00770.4.4	Pointing & Plastering	WBS	0	Design and Engineering	Engineers	Active
CN00770.4.5	Painting	WBS	0	Design and Engineering	Engineers	Active
CN00770.4.6	Waterproofing	WBS	0	Design and Engineering	Engineers	Active

*Figure 3-29 The WBS elements added under the **Banquet Hall Construction** project node*

Assigning Calendar 2 to Project

1. Choose the **Projects** option from the **Enterprise** menu in the menubar; the **Projects** window is displayed.

2. In this window, select the **Banquet Hall Construction** project and then choose the **Defaults** tab from the **Project Details** table.

3. In this tab, choose the Browse button adjacent to the **Calendar** edit box; the **Select Default Project Calendar** dialog box is displayed.

4. In this dialog box, click at the **Display: Global calendars** area; a flyout is displayed.

5. Choose the **Project Calendars** option from the displayed flyout; the created project calendars are displayed in the dialog box.

6. Select the **Calendar 2** option and then choose the **Select** button from the right side of the dialog box; the **Select Default Project Calendar** dialog box is closed.

Exporting Projects

Export projects to save the project file for future reference.

1. To export the **Banquet Hall Construction** project, ensure that the **Banquet Hall Construction** project is opened in Primavera P6.

2. Now, choose the **Export** option from the **File** menu; the **Export** wizard with the **Export Format** page is displayed.

3. In this page, ensure that the **Primavera PM/MM - (XER)** radio button is selected and then choose the **Next** button; the **Export Type** page is displayed.

4. In this page, ensure that the **Project** radio button is selected and then choose the **Next** button; the **Projects To Export** page is displayed.

5. Make sure the **Banquet Hall Construction** project is displayed in this page. Choose the **Next** button; the **File Name** page is displayed.

6. In this page, choose the Browse button adjacent to the **File Name** edit box; the **Save File** dialog box is displayed.

7. In this dialog box, browse to the *C:\PM6\c03* folder and save the file with the name **c03_CN_BNQT_tut03** and then choose the **Save** button.

8. Choose the **Finish** button; the **Primavera P6** message box is displayed with the message that the export was successful.

9. Choose the **OK** button; the file is exported.

Self-Evaluation Test

Answer the following questions and then compare them to those given at the end of this chapter:

1. Which of the following calendar pools contains calendars that can be applied to all projects?

   (a) **Global**   (b) **Project**
   (c) **Resource**   (d) **Activities**

2. Which of the following buttons in the WBS toolbar is used to show or hide the **WBS Details** table in the **Work Breakdown Structure** window?

   (a) **WBS Table**   (b) **Timescale**
   (c) **Show/Hide Bottom Layout**   (d) **Columns**

3. Primavera P6 consists of three types of calendars: _____, _____, and **Project**.

4. To open the **Calendars** dialog box, choose the _____ option from the **Enterprise** menu of the menubar.

5. The project or resource calendar can be converted into a _____ calendar.

6. In the _____ dialog box, you can adjust the working hours of every workday of a calendar.

7. By using calendar assignments, you can also do activity scheduling, tracking, and resource levelling. (T/F)

8. The Work Breakdown Structure for a project is divided into many sublevels known as WBS nodes which are also known as WBS codes. (T/F)

9. In Primavera P6, you can also calculate the performance percent complete by using the WBS milestones. (T/F)

10. The **Global** and **Resource** calendars can be modified and accessed even when no project is opened. (T/F)

Review Questions

Answer the following questions:

1. Which of the following dialog boxes is used to assign WBS category values to the desired WBS elements in the **Work Breakdown Structure** window?

   (a) **Columns**   (b) **Bars**
   (c) **Timescale**   (d) **Calendars**

2. Which of the following buttons in the WBS toolbar enables you to view the WBS in a tabular form?

   (a) **Gantt Chart**   (b) **Chart View**
   (c) **WBS Table**   (d) **Columns**

3. Which of the following dialog boxes enables to display the desired columns in the **Work Breakdown Structure** window?

   (a) **Timescale**   (b) **Bars**
   (c) **Columns**   (d) none of the above

4. Use the _____ window to view all the WBS elements related to the project.

5. In the _____ tab, you can specify the desired position and database field to the selected bar's label.

6. Technically, _____ is defined as the hierarchical arrangement of products and services produced during and by a project.

7. The **Project Calendar** pool consists of a separate pool of calendars for each_____.

8. The **ETC = remaining cost for activity** technique for calculating ETC does not factor EV (earned value) in calculating the ETC. (T/F)

9. When you are creating a new **Project** or **Resource** calendar, you also have an option to copy the calendar holidays of a **Global** calendar to new calendar. (T/F)

10. The project is the root node for the Work Breakdown Structure and has direct relationship with each WBS element. (T/F)

EXERCISES
Exercise 1

Create a WBS for the Hospital Building Project, as shown in Figure 3-30.

(Expected time: 40min)

Hints:
1. Start Primavera P6 and open the **Projects** window.
2. Open the **Hospital Building** project.
3. Open the **Work Breakdown Structure** window.
4. Create WBS.
5. Export a project.
6. Save the file with a name *c03_CONS_HOSP_ex01*.

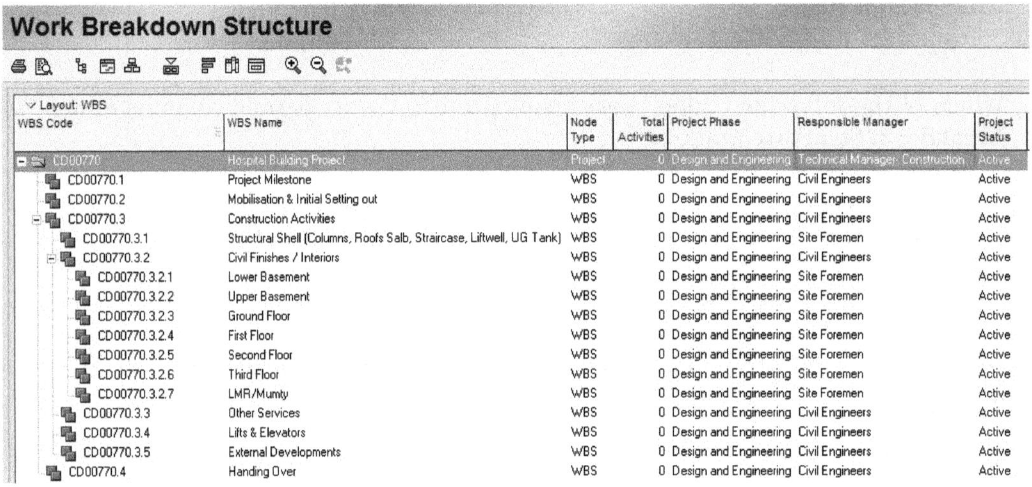

Work Breakdown Structure

⌄ Layout: WBS

WBS Code	WBS Name	Node Type	Total Activities	Project Phase	Responsible Manager	Project Status
CD00770	Hospital Building Project	Project	0	Design and Engineering	Technical Manager- Construction	Active
CD00770.1	Project Milestone	WBS	0	Design and Engineering	Civil Engineers	Active
CD00770.2	Mobilisation & Initial Setting out	WBS	0	Design and Engineering	Civil Engineers	Active
CD00770.3	Construction Activities	WBS	0	Design and Engineering	Civil Engineers	Active
CD00770.3.1	Structural Shell (Columns, Roofs Salb, Straircase, Liftwell, UG Tank)	WBS	0	Design and Engineering	Site Foremen	Active
CD00770.3.2	Civil Finishes / Interiors	WBS	0	Design and Engineering	Civil Engineers	Active
CD00770.3.2.1	Lower Basement	WBS	0	Design and Engineering	Site Foremen	Active
CD00770.3.2.2	Upper Basement	WBS	0	Design and Engineering	Site Foremen	Active
CD00770.3.2.3	Ground Floor	WBS	0	Design and Engineering	Site Foremen	Active
CD00770.3.2.4	First Floor	WBS	0	Design and Engineering	Site Foremen	Active
CD00770.3.2.5	Second Floor	WBS	0	Design and Engineering	Site Foremen	Active
CD00770.3.2.6	Third Floor	WBS	0	Design and Engineering	Site Foremen	Active
CD00770.3.2.7	LMR/Mumty	WBS	0	Design and Engineering	Site Foremen	Active
CD00770.3.3	Other Services	WBS	0	Design and Engineering	Civil Engineers	Active
CD00770.3.4	Lifts & Elevators	WBS	0	Design and Engineering	Civil Engineers	Active
CD00770.3.5	External Developments	WBS	0	Design and Engineering	Civil Engineers	Active
CD00770.4	Handing Over	WBS	0	Design and Engineering	Civil Engineers	Active

*Figure 3-30 The WBS elements added under the **Hospital Building Project** project node*

Exercise 2

Create calendars with the constraints listed below-

1. Calendar 1.1
    Working hour-   7:00 - 16:00 (9 work hours/day)
    Lunch hour-   cell no. 12 to be kept as lunch which will be counted as non-working.
    Saturday and Tuesday as weekly off.

    Calendar 2.2
    Working hour-   24x6 hours shift.
    Lunch hour-   cell no. 4,12 and 20 kept as lunch which will be counted as non-working.
    Weekly off-   Sunday

Assign **Calendar 1.1** to the project created in exercise 1.   **(Expected time: 45 min)**

Hints:
1. Login to Primavera P6 and open the **Projects** window.
2. Open the **Hospital Building** project.
3. Create calendar.
4. Export a project.
5. Save the file with a name *c03_CONS_HOSP_ex02*.

Answers to Self-Evaluation Test

1. a, **2.** c, **3. Global, Resource, 4. Calendars, 5.** global, **6. Calendar Weekly Hours, 7.** T, **8.** T, **9.** T, **10.** F

Chapter 4

Working with Activities and Establishing Relationships

Learning Objectives

After completing this chapter, you will be able to:
- *Add activities to a project*
- *Understand the Activity Details table*
- *Learn about the Activity Schedule Information*
- *Establish relationships between activities*
- *Trace logic*
- *Schedule projects*

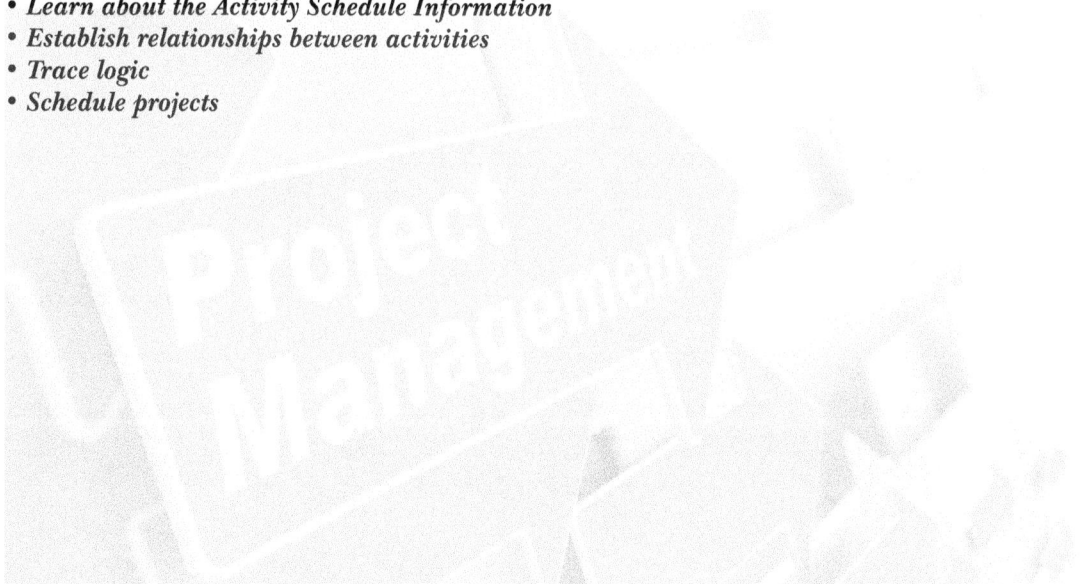

INTRODUCTION

In the previous chapter, you learned how to create the Work Breakdown Structure for a project. Also, you learned to create different types of calendars in Primavera P6 along with the procedure of creating and modifying them. In this chapter, you will learn to add activities under a WBS for a project. You will also understand the features and details of the Activities layout and establish relationship between two or more activities.

WORKING WITH ACTIVITIES

An activity is an elementary work element in a project. Activities represent the work that must take place in a specified time period. They are done at the lowest level of a WBS and are the smallest subdivision of a project. In other words, activities are small components of a project work and represent the work necessary to complete the project.

In a project, activities provide a basis for estimating, scheduling, executing, monitoring and controlling the project work. In Primavera P6, activities are also created by identifying the specific actions that need to be performed to deliver the project.

Activities represent the work that must be carried out in specific time period. To work with activities, you need to understand the options in the **Activities Layout** and **Activities Toolbar**. These are discussed next.

Activity Toolbar

The **Activity** toolbar, as shown in Figure 4-1 is displayed at the top of the **Activity** window. It contains various buttons which enable you to access the frequently used commands with ease. The buttons in the **Activity** toolbar are also used to control the display of the **Activity** window.

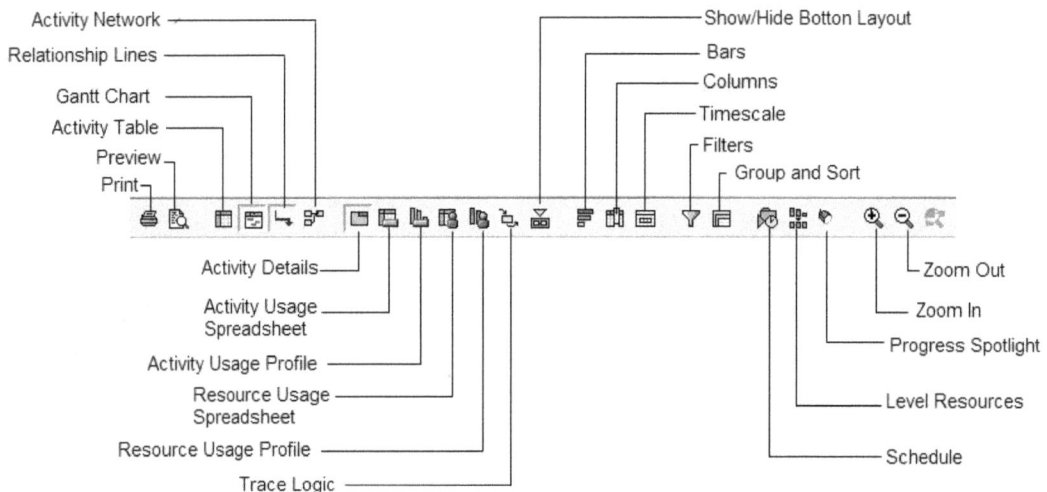

Figure 4-1 *The Activity toolbar*

Activities Layout

To create an activity layout, you need to assign the layout to be displayed in the **Activities** window. To define the activity layout, choose **Layout > Open** from **Layout: Classic WBS Layout** drop-down located below the **Activity** toolbar; the **Primavera P6** message box will be displayed. Choose the **Yes** button from the message box; the **Open Layout** dialog box will be displayed. In this dialog box, select the type of layout you want to display for the activity.

To customize the existing activity layout, choose the **Columns** button from the **Activity** toolbar; the **Columns** dialog box will be displayed. In this dialog box, select the required options from the **Available Options** area and shift them to the **Selected Options** area using the **Add to List** button.

ADDING ACTIVITIES

Activities is a set of work that must be performed to complete the project. Various functions of the project will be assigned as activities that are needed to be carried out in a given time. You can create activities for a project in the **Activities** window of the Primavera P6. To do so, choose and open the required project from the **Projects** window; the **Activities** window will be displayed, as shown in Figure 4-2.

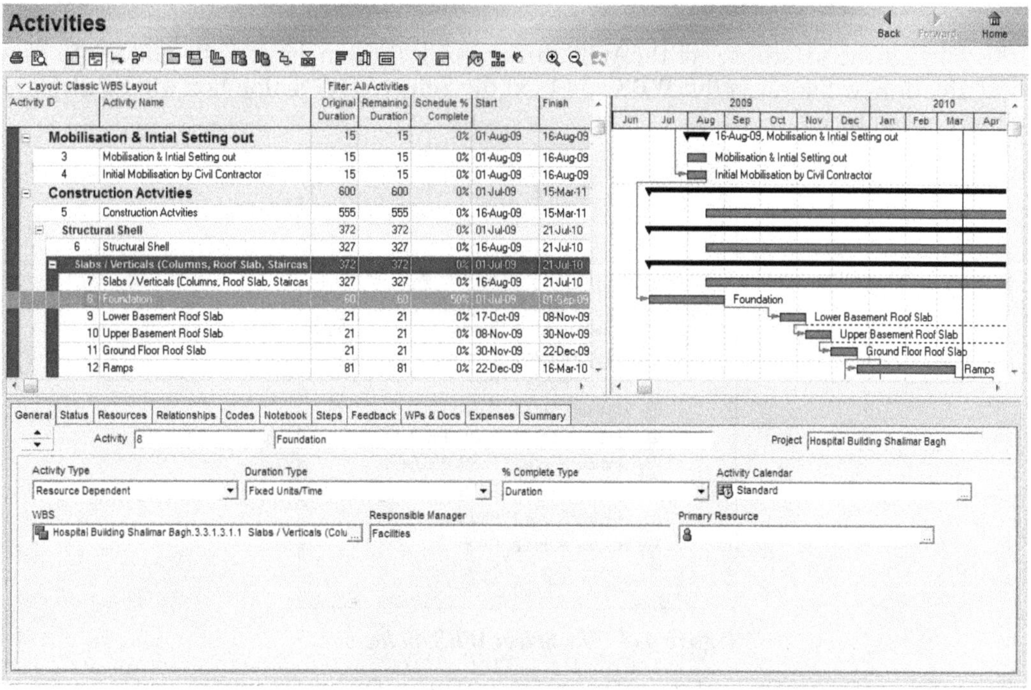

*Figure 4-2 The **Activity** window with upper and lower layouts*

In this window, choose the **Add** button from the Command bar; the **New Activity** wizard with the **Activity Name** page will be displayed. In this wizard, the activity id is entered by default, enter the name of activity in the **Activity Name** edit box. Now, choose the **Next** button; the **Work Breakdown Structure** page will be displayed, as shown in Figure 4-3.

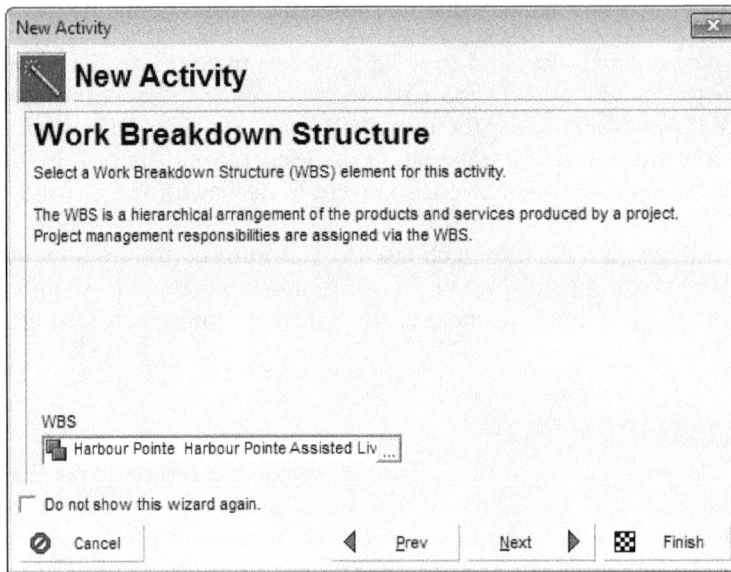

*Figure 4-3 The **New Activity** wizard with the **Work Breakdown Structure** page*

In this page of the wizard, select the WBS under which the activity is to be added. To do so, choose the Browse button in the **WBS** edit box; the **Select WBS** dialog box will be displayed, as shown in Figure 4-4.

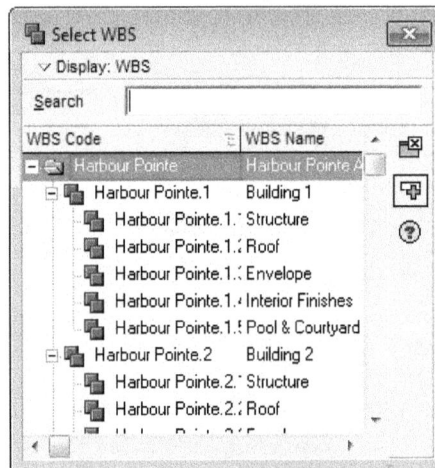

*Figure 4-4 The **Select WBS** dialog box*

In this dialog box, select the required WBS from the **WBS Name** list and then choose the **Select** button; the WBS will be assigned and the dialog box will be closed. Choose the **Next** button; the **Activity Type** page will be displayed, as shown in Figure 4-5.

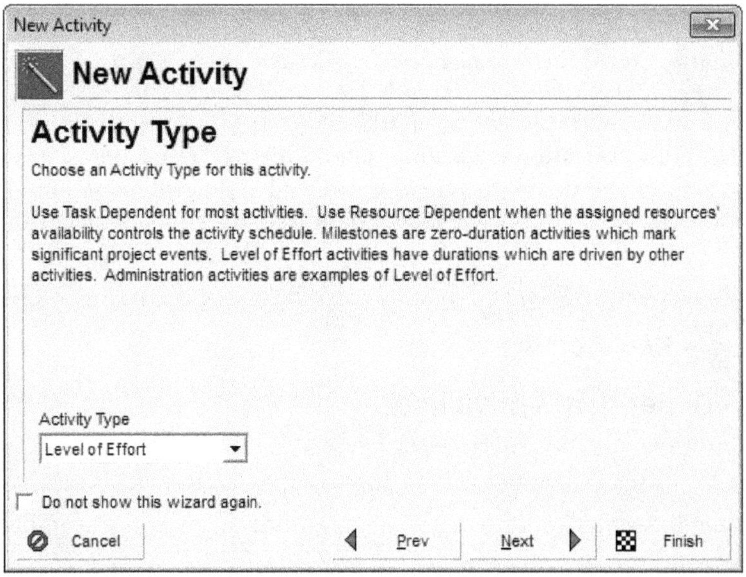

*Figure 4-5 The **New Activity** dialog box with the **Activity Type** page*

In this page, select the required activity type from the **Activity Type** drop-down list and then choose the **Next** button; the **Assign Resources** page will be displayed. In this page, assign the resources required to complete the activities and then choose the **Next** button; the **Duration Type** page will be displayed. In this page, you need to select the duration type option from the **Duration Type** drop-down list according to it the duration will be defined. Choose the **Next** button; the **Activity Units and Duration** page will be displayed, as shown in Figure 4-6.

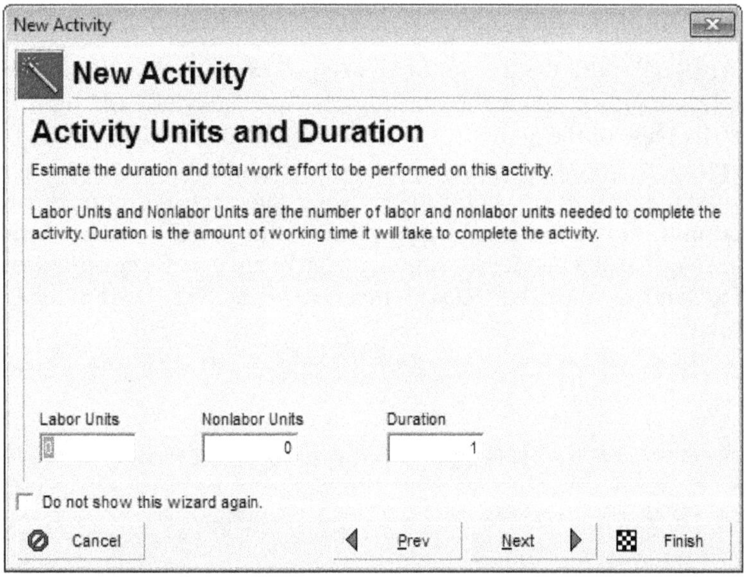

*Figure 4-6 The **New Activity** window with the **Activity Units and Duration** page*

Note
*On selecting the **Start Milestone** and **Finish Milestone** option from the **Activity Type** drop-down list, the **Dependent Activities** page will be the next displayed page.*

In the **Activity Units and Duration** page, enter labor units and non labor units that are required to complete that activity and then enter the duration value that will be required for the activity completion. Choose the **Next** button; the **Dependent Activities** page will be displayed, as shown in Figure 4-7.

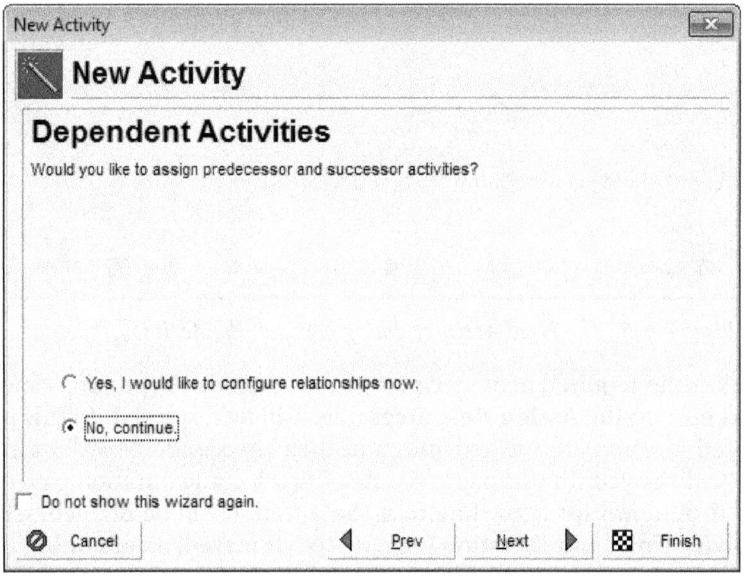

*Figure 4-7 The **New Activity** wizard with the **Dependent Activities** page*

In this page, you can select the **Yes, I would like to configure relationships now** radio button to assign the predecessor and successor relationships between activities and then you will progress to the **More Details** page of the activity. If you select the **No, Continue** radio button and then choose the **Next** button; the **More Details** page will be displayed, as shown in Figure 4-8, and will not allow to assign relationship. In this page, if you select the **Yes, continue configuring the activity information** radio button it will allow you to configure more information of activity in terms of expenses, and activity codes and then you will move to **Congratulations** page. But, if you select the **No, Continue** radio button and then choose the **Next** button; the **Congratulations** page will be displayed.

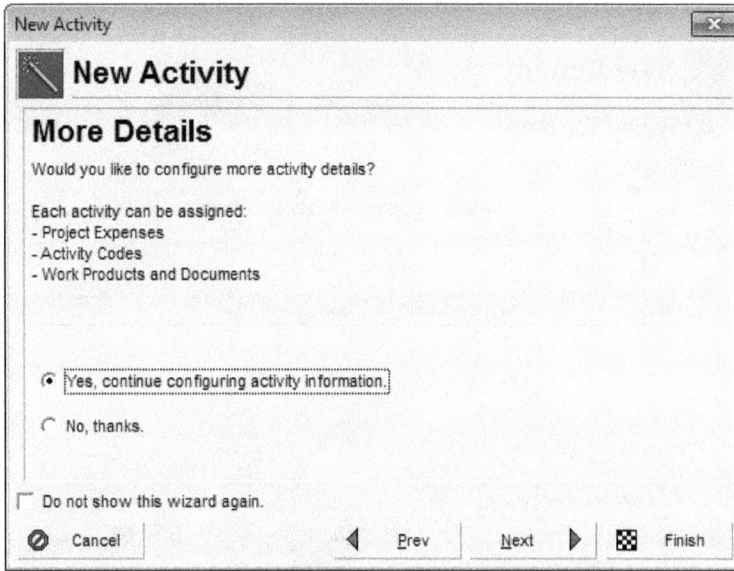

Figure 4-8 The New Activity wizard with the More Details page

Select the **Yes, continue configuring the activity information** radio button and then choose the **Next** button; the **Expenses** page will be displayed, as shown in Figure 4-9. In this page, choose the **Add** button; a row will be added. Enter the expense item and its cost values under the **Expense item** and **Budgeted Cost** columns, respectively. Now, choose the **Next** button; the **Activity Codes** page will be displayed, as shown in Figure 4-10.

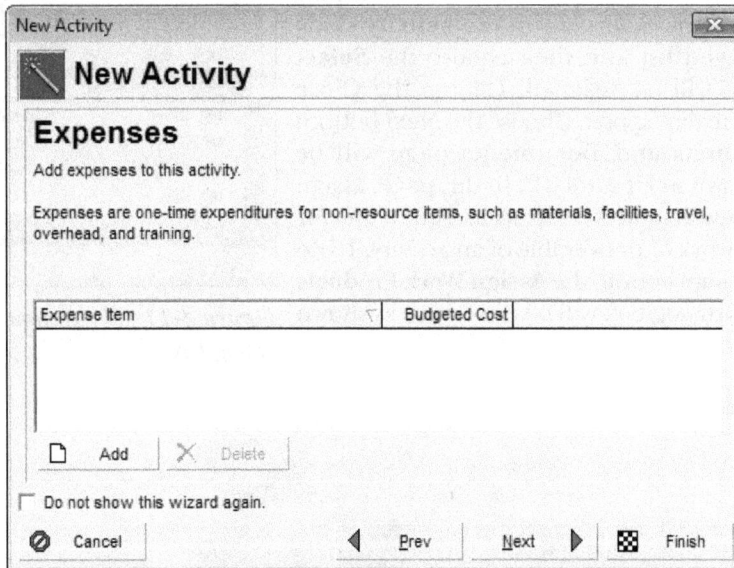

Figure 4-9 The New Activity wizard with the Expenses page

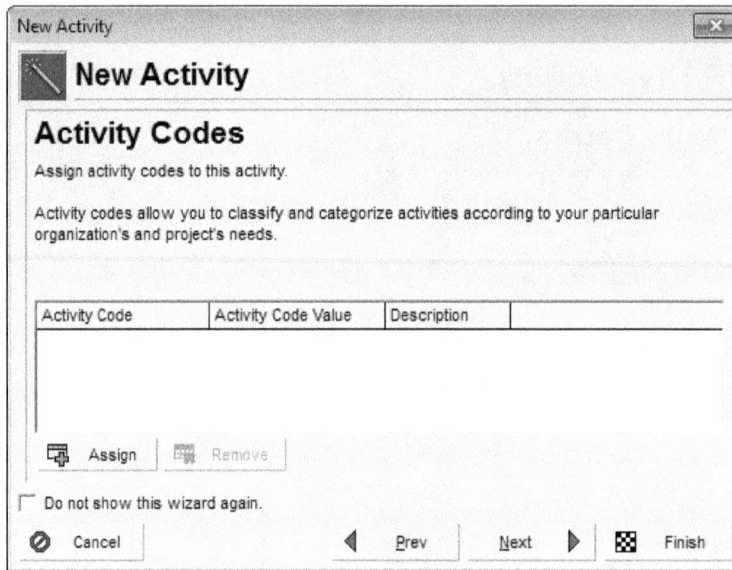

Figure 4-10 *The **New Activity** wizard with the **Activity Codes** page*

In this page, enter the code value for activity which helps you to classify and categorize activities according to particular organization's and project's needs. Choose the **Assign** button; the **Assign Activity Codes** dialog box will be displayed, as shown in Figure 4-11. In this dialog box, select the required radio button to display the codes accordingly. Select any of the **Activity Code** from the displayed list and then choose the **Select** button the code will be assigned. Choose the **Close** button to close the dialog box. Choose the **Next** button; the **Work Products and Documents** page will be displayed, as shown in Figure 4-12. In this page, assign the reference document which will act as a guideline for performing the work or deliverable of an activity. To do so, choose the **Assign** button; the **Assign Work Products and Documents** dialog box will be displayed, as shown in Figure 4-13.

Figure 4-11 *The **Assign Activity Codes** dialog box*

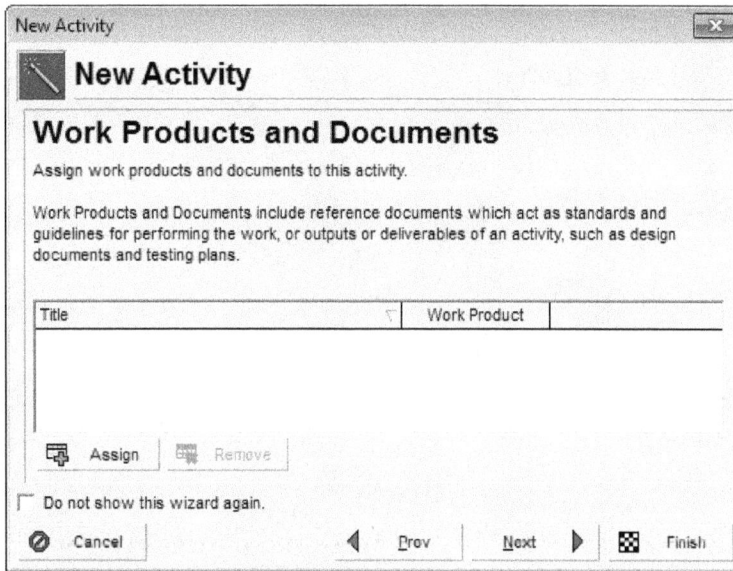

Figure 4-12 The New Activity wizard with the Work Products and Documents page

In this dialog box, ensure that the **All WPs & Docs** option is selected in the **Display** drop-down list. Select the required option from the list and then choose the **Select** button; the selected document will be assigned. Now, choose the **Close** button to close the dialog box. Now, choose the next button; the **Congratulations** page will be displayed, as shown in Figure 4-14. This page will inform you that the activity is created successfully.

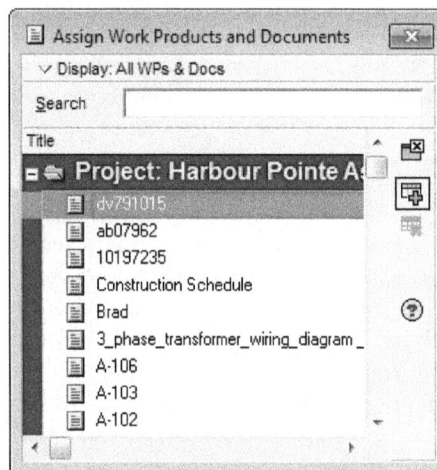

Figure 4-13 The Assign Work Products and Documents dialog box

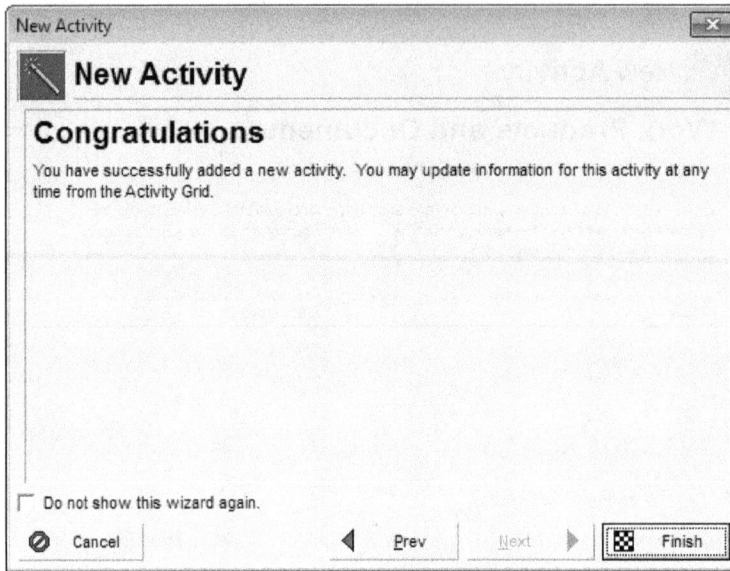

Figure 4-14 *The* **New Activity** *wizard with the* **Congratulations** *page*

The **Activities** window is divided into two main layouts: upper and lower. The upper layout shows the activity table, Gantt chart and activity network. The lower layout mainly comprises of the **Activity Details** table which is described next.

In the **Activities** window, the **Activity Details** table is displayed at the bottom. It helps to add the details to the activities in the project. Using this details table, you can define different types of information associated with an activity. With the help of these information, you can identify components associated with each activity such as the activity ID and name, activity codes, predecessor and successor activities, activity start and finish dates, WBS element, roles and resources, activity type, duration type, percent complete type, and so on.

ACTIVITY DETAILS TABLE

In the **Activity Details** table, you can provide general information for the selected activity such as the activity type, duration type, WBS assignment, primary resource, and activity calendar. To display the **Activity Details** table, choose **Show on Bottom > Activity Details** from the **Layout** drop-down list; the **Activity Details** table will be displayed, as shown in Figure 4-15. You can customize the tabs of the **Activity Details** table. To do so, right-click on the space adjacent to the tabs in this table; a shortcut menu will be displayed. Choose the **Customize Activity Details** option; the **Activity Details** dialog box will be displayed, as shown in Figure 4-16.

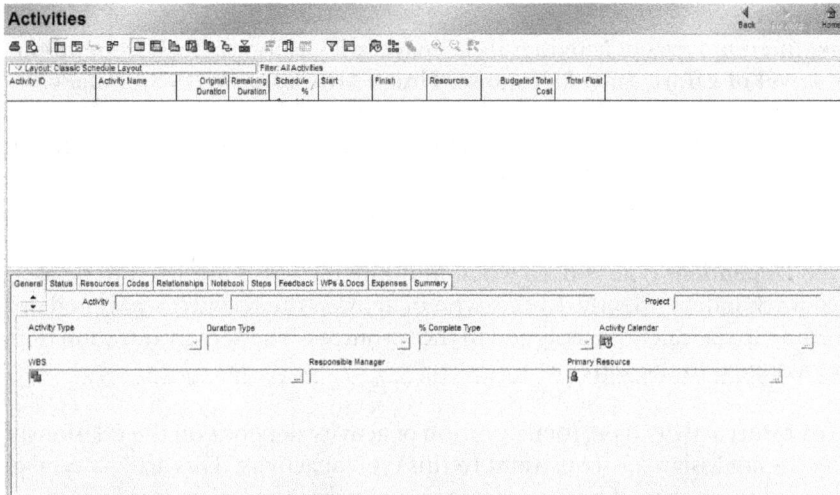

*Figure 4-15 The **Activities** window with the **Activity Details** table*

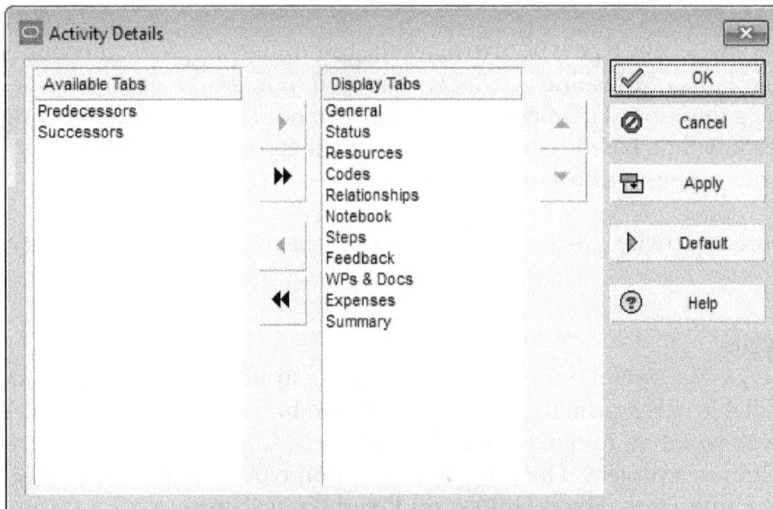

*Figure 4-16 The **Activity Details** dialog box*

In this dialog box, add those tabs in the **Display Tabs** area that you want to display in the **Activity Details** table. Now, choose the **OK** button; the **Activity Details** dialog box will be closed. The tabs of the **Activity Details** table are described later in this chapter.

General Tab

The **General** tab helps to view and edit general information of the selected activity. This tab contains details such as ID, name, type, activity duration type, status, responsible manager, % complete, activity calendar and so on. Using this tab, you can also view or edit the project ID and the WBS under which the activity is assigned. The fields of the **General** tab are described next.

Activity Types

There are six different types of activities in Primavera P6 as listed: **Task Dependent, Resource Dependent, Level of Effort, Start Milestone, Finish Milestone, WBS Summary**.

The **Task Dependent** activity type is used when multiple resources assigned to the same activity need to work together. The **Task Dependent** activity enables scheduling resources according to the calendar assigned to an activity rather than the resource calendar.

The **Resource Dependent** type is used when multiple resources are need to be assigned to an activity which work independently. In this type of activity, the resources assigned are supposed to work according to the calendar assigned to the resources. The activity duration depends upon the resources assigned to the activity.

In the **Level of Effort** activity type, total duration of activity depends on the relationship between activities. You cannot assign any constraint to this type of activity. This activity type gets its start date from the early start date of its predecessor activity and the finish date from the early finish date of its successor activity. Then its own calendar will calculate the duration between the start and finish dates.

The **Start Milestone** activity type is used to mark the beginning of a phase or communicate project deliverable. The **Finish Milestone** activity type is used to mark the end phase of a project and it does not have a duration. These two activity types do not have the resource assignments, role assignments, or any time based cost such as labor or material. In fact, they may have primary resources and the expenses associated with it.

The **WBS Summary** activity type is used when a group of activities come under a common WBS level.

Duration Type

The duration type is assigned only when resources are available in an activity. Different types of duration available in the primavera module are used for scheduling a project. You can select the duration type based on resources, schedule, or costs. Cost is the most important factor for updating activities in a project. There are four duration types **Fixed Duration & Units, Fixed Durations and Units/Time, Fixed Units**, and **Fixed Units/Time**.

The **Fixed Duration & Units** type is used for Task Dependent activities. This is the default duration type selected in the **New Activity** dialog box. This duration type is used when duration and total resources for the activity is fixed. For example, if an activity requires 15 days and 60 labor units as resources to complete, then primavera automatically calculates Units/Time for (60/15= 4 units/day).

The **Fixed Durations and Units/Time** is also used for Task Dependent activities. This duration type is used when a minimum number of resources are required per time period for an activity. This duration type co-relates with the resources and duration and according the changes occur in both when the original duration is changed. For example, if an activity requires one excavator and two operators to operate then it will take the required time. If you need the job work to be done before time then on increasing the number of operators will not help and there will be requirement to increase resources such as excavator.

The **Fixed Units** type is mostly used for the Resource Dependent activities. This type is assigned when the total amount of work is constant and the resources are used based on time. For example, to excavate 60 cubic meters in 15 days you need 4 cubic meters to be excavated each day. If you want to complete the same job in 6 days then you need to excavate 10 cubic meters each day.

The **Fixed Units/Time** type is also used for Resource Dependent activities. This type is used when the time of work will be affected by the resources.

Activity Calendar

In this edit box, you need to assign the dates calendar for the assigned activity. To do so, choose the Browse button from the **Activity Calendar** edit box; the **Select Activity Calendar** dialog box will be displayed. Select the activity calendar from this dialog box. In case you need the information other than the information specified in a predefined calendar, then you can create you own calendar.

WBS

You can assign a WBS to different activities. To do so, choose the Browse button on the right of the **WBS** edit box in the **General** tab of the **Activity Details** table; the **Select WBS** dialog box will be displayed. In this dialog box, select the WBS under which you want to create the activity.

Responsible Manager

This edit box of the **Activity Details** table displays the name of the responsible manager that is assigned with the selected activity's WBS elements.

Primary Resource

This edit box displays the name of the selected activity's primary resource. The primary resource is the person responsible for the overall work related to the activity and for updating its status. To assign a primary resource, choose the Browse button available on the right of the **Primary Resource** edit box; the **Select Primary Resource** dialog box will be displayed. In this dialog box, select the required option of the primary resource and then choose the **Select** button; the resources will be assigned and the dialog box will be closed.

Status Tab

The **Status** tab enables you to view and edit detailed schedule information of the selected activity, as shown in Figure 4-17. The status information includes the status of activity, duration type, total float and free float, activity constraints and labor units. In this tab, you can also specify the anticipated start and finish dates of the selected activity. The Primavera P6 module automatically recalculates the time value and period that you enter according to the project's calendar and the standard time period defined by the network administrator.

To view the available time period abbreviations, choose the **Admin Preferences** option from the **Admin** tab; the **Admin Preferences** dialog box will be displayed. Choose the **Time Periods** tab and enter the required values in the edit boxes displayed. You must have appropriate access rights to edit admin preferences. The other fields in the **Status** tab are described next.

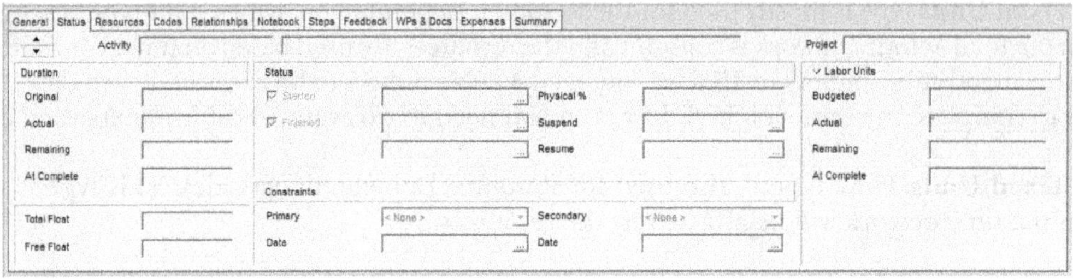

Figure 4-17 *The **Activity Details** table with the **Status** tab chosen*

Duration

In the **Duration** area, you can add or update the duration of an activity. To add the original activity duration, enter a value in the **Original** edit box. Enter the actual start time of the activity in the **Actual** edit box. Enter the remaining duration for the activity in the **Remaining** edit box. Enter a new completion estimate value of assigned activity in the **At Complete** edit box; the duration of completion time for the selected activity will be calculated. If the selected activity is in progress, type a new at completion estimate (At Complete Duration = Actual Duration + Remaining Duration).

The total float is the amount of time for which the selected activity can be delayed without delaying the project's finish date. You can enter its value in the **Total Float** edit box. Free float is the amount of time the selected activity can be delayed without delaying the immediate successor activities.

Status

In the **Status** area, you need to update the start and finish date of an activity by selecting the corresponding check boxes. When an activity is in progress, you are required to update the status of an activity.

If the activity has started, select the **Started** check box and then specify the actual start date in the field corresponding to the **Started** check box by using the Browse button. To enter the project's finish date, select the **Finished** check box and then specify the value in the edit box corresponding to the **Finished** check box. Once the activity has started, you can enter the expected end date in the **Exp Finish** edit box. Anyone who has access to the project can edit this date.

You can enter the activity completion percent **Duration %** edit box. The duration % is calculated from the original and remaining durations.

Constraints

Sometimes activity must be accomplished according to specific dates rather than the dates determined by other activities in the project. Constraints are used only when the activities must start or finish on a specific date. To signify dependence on specific dates, assign primary and secondary constraints to activities.

To assign a constraint, select the desired option from the **Primary** drop-down list in the **Constraints** area of the **Status** tab. The secondary constraint will be available only when primary constraint has been assigned. If necessary, you can assign the secondary constraint by selecting the required option from the **Secondary** drop-down list.

Labor/Nonlabor Units/Cost or Material Cost
In the **Labor Units** area, you can change the name of the area by clicking on the **Labor Units** drop-down and select either the labor/nonlabor units/cost, or material cost. After selecting the desired option from the **Labor Units** drop-down, you need to enter the **Budgeted**, **Actual**, **Remaining** and **At Complete** values in their corresponding value fields in terms of numbers and currency.

Resources Tab
The **Resources** tab helps to specify the resources essential for the project and also the project level resource permissions required for the application of time sheet. These permissions allow the resources to be assigned to activities, and generate report when activities are completed.

Relationships Tab
This tab is used to establish a relationship between activities. In this tab, the **Predecessors** and **Successors** areas will be displayed. To assign relation to either the successor or predecessor, choose the **Assign** button from the respective area. On choosing the **Assign** button from the **Predecessor** area; the **Assign Predecessor** dialog box will be displayed. Similarly, on choosing the **Assign** button from the **Successors** area; the **Assign Successor** dialog box will be displayed. In this dialog box, select the activity and then choose the **Assign** button; the activity will be added to the selected area. Choose the **Close** button to exit the dialog box.

Codes Tab
In this tab, you can assign a code and value to the selected activity. If an applicable activity code or value does not exist, then choose the **Assign** button from the **Codes** tab; the **Assign Activity Codes** dialog box will be displayed. In this dialog box, select the radio button corresponding to the activity code type that you want to assign.

Notebook Tab
The **Notebook** tab helps in assigning notebook topics, details, and description to the selected node or project.

Steps Tab
In this tab, you can assign various steps that are to be executed in a project. You can also maintain the record of the steps, whether they have completed or are pending.

Expenses Tab
This tab enables you to add the expenses of a project or an activity. To add the expenses of a particular activity, select the activity from the **Activity** window and then add the expense item of that particular activity using the **Add** button. On doing so, the expense item will be added under that activity. You can edit its name by double-clicking in the **Expense item** column. Similarly, you can edit other entries by double-clicking in the respective columns.

Summary Tab

In this tab, you can add the summary of the activities of a project. This tab enables you to add the labor/nonlabor units and duration involved in an activity. You can fill the information like the budgeted labor/nonlabor units, and actual, remaining, and % completion of activities in their respective edit boxes.

ESTABLISHING RELATIONSHIP BETWEEN ACTIVITIES

In Primavera P6, you need to establish a relationship between activities to schedule the project. On establishing the relationship, a link between the successor and predecessor is made. This relationship indicates that any activity will begin only after first activity is completed. Once the relationship is assigned, schedule the project to calculate the early and late start and end dates for each activity.

Generally, relationships are established between activities in the same project. However, you can define activity relationships in any other project, even if the project is not opened in the current display. There are several methods for assigning relationships. Use the Activity Network to visualize the flow of logic as you link activities, or use the Gantt Chart to view relationships according to time. You can choose the **Relationships** tab in the **Activity Details** table to assign relationships to the activities in the same project or in other projects in the EPS.

Relationship Types

There are four types of relationships: FS, FF, SS, SF. These relationship types are assigned from predecessor activity to the successor activity. The activity relationship types are described next.

Relationship Types	Symbols	Description
1. **(FS) Finish to start**		It indicates that the successor activity can begin only when the predecessor activity has completed.
2. **(FF) Finish to finish**		It indicates that the finish of the successor activity depends on the finish of the predecessor activity
3. **(SS) Start to start**		It indicates that the start of the successor activity depends on the start of the predecessor activity
4. **(SF) Start to finish**		It indicates that the successor activity cannot finish until the predecessor activity starts.

A successor activity cannot start or finish until its predecessor activity starts or finishes. This develops a lag for the relationship.

Lag

Lag is defined as the number of time units consumed from the start or finish of an activity to the start or finish of its successor. Lag can be a positive or a negative value.

To enter a value for the lag, you need to select a calendar. This calendar will help you to calculate the lag between the predecessors and successors for all activities. If the calendar is not selected, then the successor activity calendar is used to calculate the lag.

To select the calendar for relationship lag, choose the **Schedule** option from the **Tools** menu; the **Schedule** dialog box will be displayed. In this dialog box, choose the **Options** button; the **Schedule Options** dialog box will be displayed, refer to Figure 4-20. In this dialog box, the **Calculate start-to-start lag from** area will help you to set options for the lag calculation. The other options of this dialog box are described later in this chapter.

Assigning Relationships

In the **Activity Details** table, select the activity to which you want to add a predecessor or successor relationship. To assign a relationship, choose the **Assign** button in either the predecessor or the successor area. On choosing the **Assign** button, the **Assign Predecessors** dialog box will be displayed, showing the list of activities. Select an activity from the displayed list and then choose the **Assign** button. The activity is assigned either to the predecessor or successor relationship. Choose the **Close** button. After placing the activities in predecessor or successor area, you need to assign the relationship type. To do so, double-click on the **Relations** field; a drop-down list will be displayed. Select an option from the list to assign the relationship. If there is any lag value, you can enter the lag value in the **Lag** field.

> **Tip**
> *You can use the **Predecessors** or **Successors** tab in the **Activity Details** table to assign relationships. The **Relationships** tab combines the predecessor and successor information in a single tab. The data thus stored in the **Relationships** tab is synchronized with the **Predecessors** and **Successors** tabs.*

TRACE LOGIC

The Trace Logic enables you to examine the path after assigning a relationship. The Trace Logic chart displays the chart details of a project that is assigned to a predecessor or successor relationship. To display the Trace Logic chart, click on the **Layout Options** bar; a menu will displayed. In this menu, choose the **Show on Bottom > Trace Logic** option; the Trace Logic chart will be displayed with the details of relationship activities, as shown in Figure 4-18.

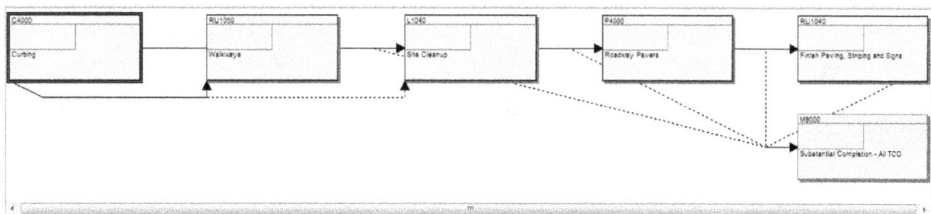

*Figure 4-18 The **Trace Logic** diagram*

SCHEDULING

Scheduling of a project is done to examine all the activities in a network from their start to end. For calculating project schedules, the Critical Path Method (CPM) scheduling technique is used. In this method, activity durations and relationships between activities are used to calculate the project schedule. To calculate the schedule, choose the **Schedule** option from the **Tools** menu; the **Schedule** dialog box will be displayed, as shown in Figure 4-19.

In this dialog box, you need to set the data date. The data date is a date that is used as a starting point for scheduling a project plan. Each time you update a project, you need to set the data date as it helps you in examining the progress of activities. Its use is reflected in the project when you recalculate the schedule. For example, assume that a project is updated on every Monday and the current data date of the project is Monday, 1st December. Now, as the project starts on 1st December and it is needed to be updated after a week which means on 8th December, then you can advance the data date to 7th December. The schedule is updated to examine the critical path activities and to examine the progress of activities. Therefore, it is required to update the schedule after every week.

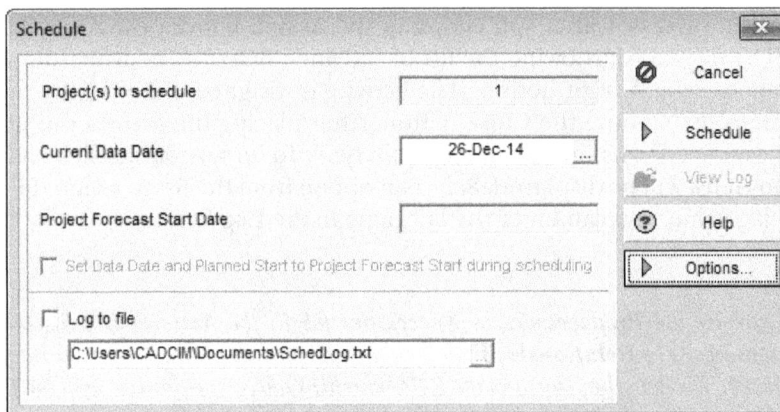

Figure 4-19 The *Schedule* dialog box

Schedule Options

To set the schedule options, choose the **Options** button from the **Schedule** dialog box; the **Schedule Options** dialog box will be displayed, as shown in Figure 4-20. In this dialog box, you can select the **Ignore relationships to and from other projects** check box to ignore the relationship of one project with other project which is being scheduled. The **Use Expected Finish Dates** check box enables you to calculate the schedule using the assigned expected finish date. Select the **Schedule automatically when a change affects dates** check box to automatically calculate the schedule each time the activity data date is changed. The **Level resources during scheduling** check box enables you to level the resources automatically while scheduling the project. The **Recalculate assignment costs after scheduling** check box helps you to calculate the cost of resources and role assignments that are assigned multiple rates.

The **When scheduling progressed activities use** area specifies the type of logic used to schedule the activities that are in progress. This area comprises of three radio buttons. The **Retained Logic** radio button enables you to schedule only if all predecessor activities are complete.

When you select the **Progress Override** radio button, the network of the activities in a project is ignored and the activity remains in progress. When you select the **Actual Dates** radio button, the backward and forward passes are scheduled using actual dates. The backward pass is used to determine the Late Start Time (LST) for each activity. The late start time represents the time at which an activity starts with delays in a project. It is a backward process in which the process starts in a reverse direction. The forward pass is a CPM technique used to determine the early start and the early finish dates of an activity.

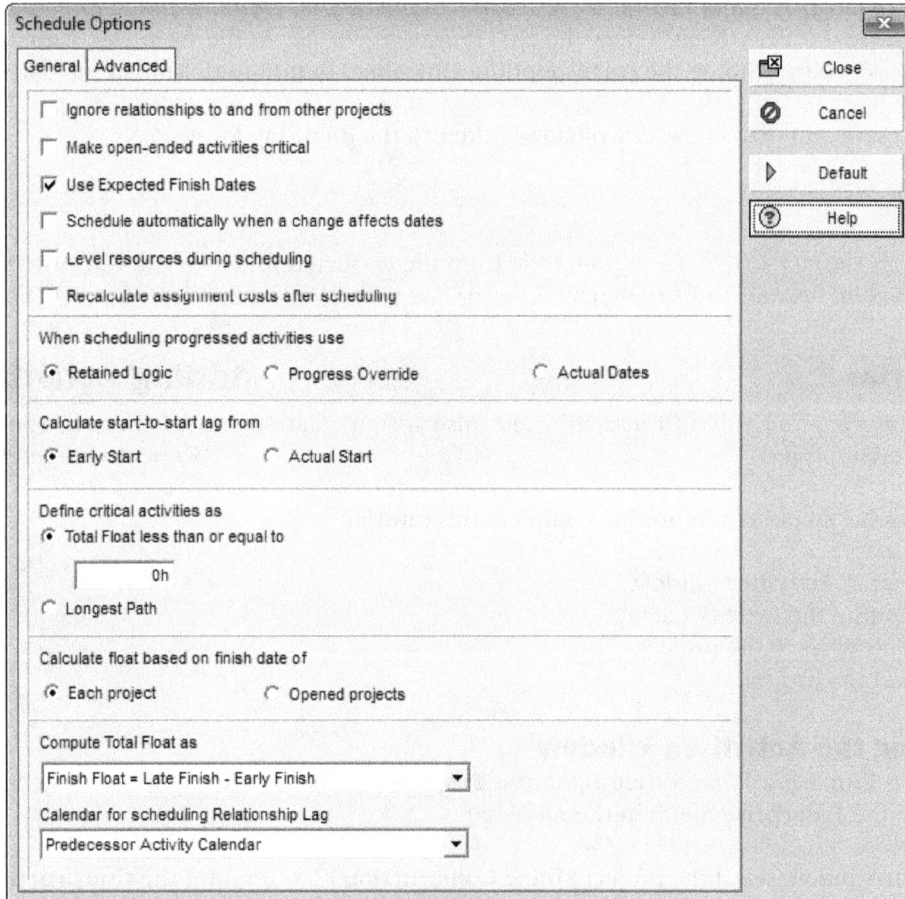

*Figure 4-20 The **Schedule Options** dialog box*

You can select the **Each project** radio button to calculate the float based on the finish date of each project in the **Calculate float based on finish date of** area. On selecting the **Opened projects** radio button, you can calculate the float of the opened projects only. The total float can be calculated by selecting any of the formulas from the **Compute Total Float as** drop-down list. The relationship lag on scheduling a project will depend upon the option selected from the **Calendar for scheduling Relationship Lag** drop-down list.

TUTORIALS

To perform the tutorials of this chapter, you need to complete Tutorial 1 and Tutorial 2 of Chapter 2.

To perform this tutorial, you need *c03_Primavera_P6_tut.zip* file. You can download this files from *www.cadcim.com* using the following steps.

1. To download the file, browse to *Textbooks > Civil/GIS> Primavera P6 > Exploring Oracle Primavera P6 v7.0*. Next, select the *c03_Primavera_P6_tut.zip* file from the **Tutorial Files** drop-down list. Choose the corresponding **Download** button to download the data file.

2. Now, save and extract the downloaded folder to the following location:

   C:/PM6

3. Import the *c03_CONS_Home_tut02* file from the extracted folder to the Construction EPS created in Tutorial 1 of Chapter 2.

Tutorial 1   Adding Activities

In this tutorial, you will add activities and also assign relationship to them in the **Home Construction** project.   **(Expected time: 40 min)**

The following steps are required to complete this tutorial:

a. Open the **Activities** window.
b. Customize the Activity Layout.
c. Add activities to the project.
d. Export the project.

Opening the Activities Window

1. Open Primavera P6 and then open the **Projects** window by choosing the **Project** option from the **Enterprise** menu in the menubar.

2. In this window, select the project **Home Construction** placed under the **Construction EPS** node and right-click; a shortcut menu is displayed.

3. Choose the **Open Project** option from this menu; the **Activities** window is displayed.

Customizing the Activity Layout

1. In the **Activities** window, choose **Layout > Open** from the **Activity Layout** drop-down; the **Primavera P6** message box is displayed prompting you to save this layout. Choose the **No** option; the **Open Layout** dialog box is displayed.

2. In this dialog box, select the **Classic WBS Layout w/ 3 line timescale** option from the **User - admin** head. Choose the **Open** button; the selected layout is displayed.

3. Choose the **Columns** button from the **Activity** toolbar; the **Columns** dialog box is displayed, as shown in Figure 4-21.

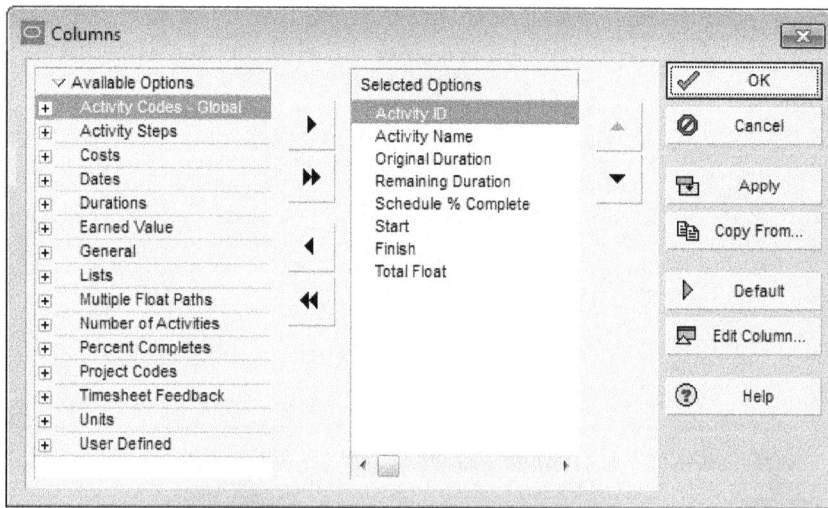

Figure 4-21 The Columns dialog box

4. In this dialog box, select the **Remaining Duration**, **Schedule % Complete**, and **Total Float** options from the **Selected Options** area.

5. Choose the **Remove from list** button; the selected options are moved to the **Available Options** area.

6. In the **Available Options** area, expand the **General** node and select the **Activity Type** option.

7. Choose the **Add to list** button; the selected option move to the **Selected Options** area.

8. Using the **Move up in list** button, arrange the options in the **Selected Options** area, as shown in Figure 4-22.

9. Choose the **OK** button to exit the **Columns** dialog box; the **Activity** window is displayed with the selected options.

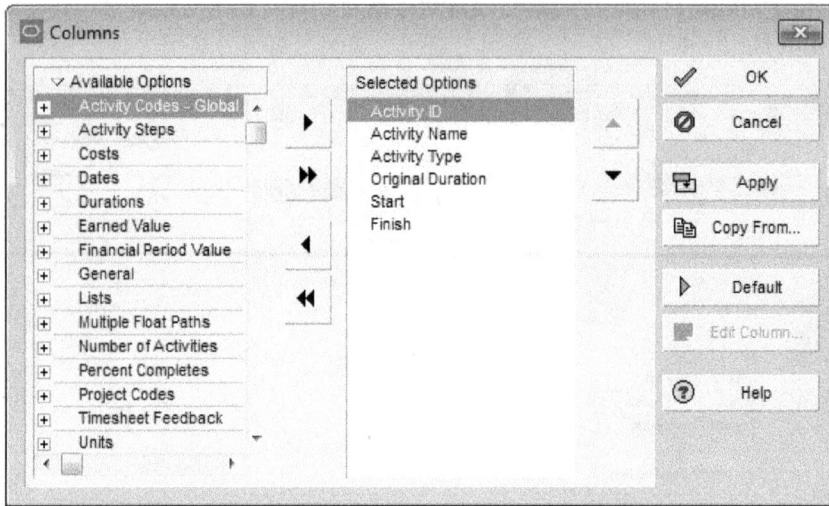

*Figure 4-22 The **Columns** dialog box with the **Selected Options** area*

Adding Activities

1. In the **Activities** window, choose the **Add** button from the Command bar; the **New Activity**
    dialog box with the **Activity Name** page is displayed, as shown in Figure 4-23.

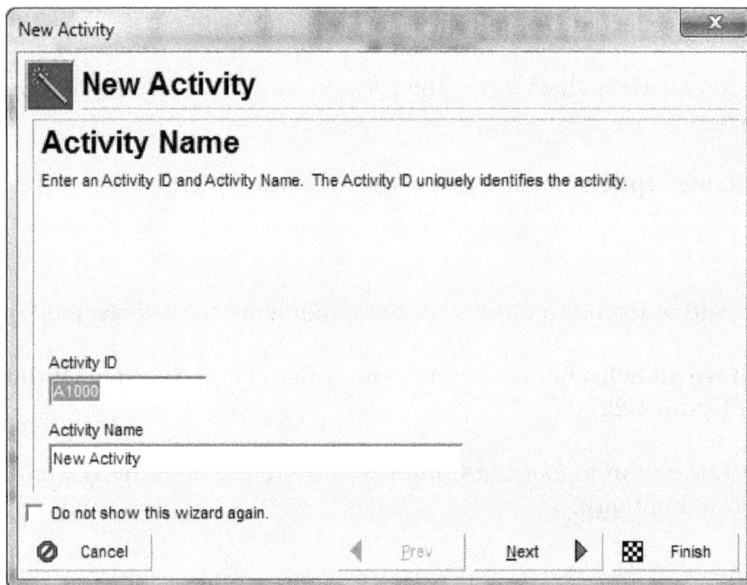

*Figure 4-23 The **New Activity** dialog box with the **Activity Name** page*

2. In this dialog box, enter **A1000** and **Project Start** in the **Activity ID** and **Activity Name**
    edit boxes, respectively.

3. Choose the **Next** button; the **Work Breakdown Structure** page is displayed.

4. In this page, specify the **Home Construction** WBS in the **WBS** edit box. Choose the **Next**
    button; the **Activity Type** page is displayed.

5. In this page, select the **Start Milestone** option from the **Activity Type** drop-down list.

6. Choose the **Next** button; the **Dependent Activities** page is displayed. In this page, ensure that the **No, continue** radio button is selected.

7. Choose the **Next** button; the **More Details** page is displayed. Ensure that the **No, thanks** radio button is selected.

8. Choose the **Next** button; the **Congratulations** page is displayed and the activity is added.

9. Choose the **Finish** button; the activity is added under the **Home Construction** WBS head.

10. To add another activity, choose the **Add** button from the Command bar; the **New Activity** dialog box with the **Activity Name** page is displayed.

11. In this page, ensure that **A1010** is entered in the **Activity ID** edit box and enter **Project Management** in the **Activity Name** edit box.

12. Choose the **Next** button; the **Work Breakdown Structure** page is displayed. In this page, specify the **Home Construction** WBS in the **WBS** edit box.

13. Choose the **Next** button; the **Activity Type** page is displayed.

14. In this page, select the **Level of Effort** option from the **Activity Type** drop-down list.

15. Choose the **Next** button twice; the **Duration Type** page is displayed. Ensure that the **Fixed Duration & Units** option is selected in the **Duration Type** drop-down list.

16. Choose the **Next** button; the **Activity Units and Duration** page is displayed. Enter **5** in the **Duration** edit box.

17. Choose the **Next** button; the **Dependent Activities** page is displayed. In this page, ensure that the **No, continue** radio button is selected.

18. Choose the **Next** radio button; the **More Details** page is displayed. Ensure that the **No, thanks** radio button is selected.

19. Choose the **Next** button; the **Congratulations** page is displayed.

20. Choose the **Finish** button; another activity is added under the **Home Construction** head. Figure 4-24 shows the **Activities** window with the added activities and Gantt chart.

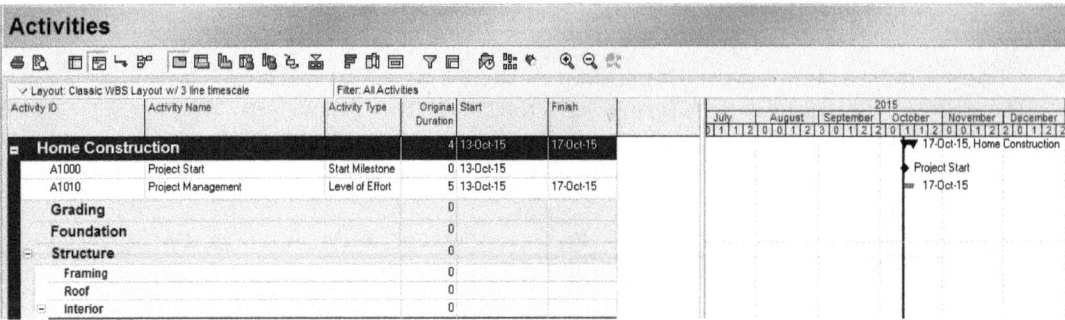

*Figure 4-24 Partial view of the **Activities** window with the added activities and Gantt chart*

21. Similarly, add another activity, as shown in Figure 4-25 in the **Activities** window with the Project Complete as the activity name and **Finish Milestone** as the activity type.

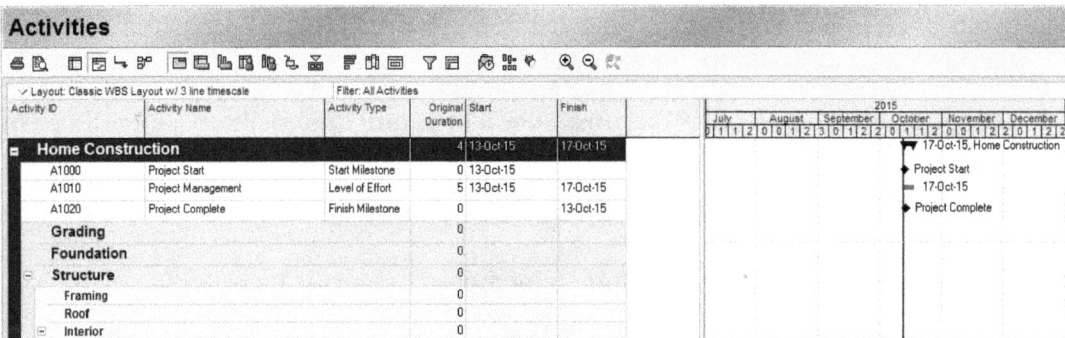

*Figure 4-25 Partial view of **Activities** window with the added activities*

22. Repeat the procedure followed in step 1 through step 9 and add other activities with the entries, as shown in Figure 4-26.

23. Select the **Fixed Duration & Units** duration type for Task Dependent activities and select **Fixed Duration and Units/Time** duration type for Resource Dependent activities.

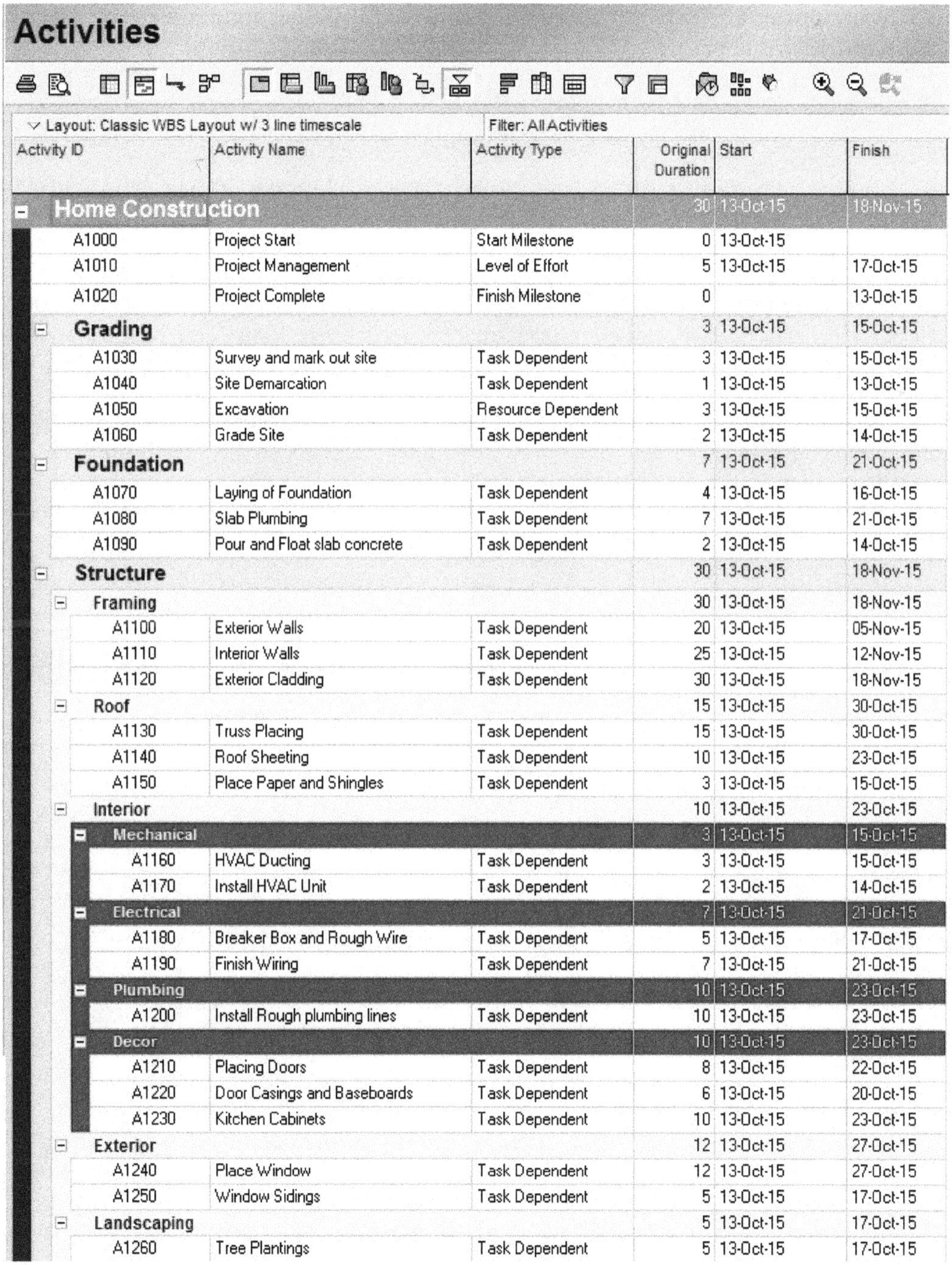

Activities

Layout: Classic WBS Layout w/ 3 line timescale — Filter: All Activities

Activity ID	Activity Name	Activity Type	Original Duration	Start	Finish
Home Construction			30	13-Oct-15	18-Nov-15
A1000	Project Start	Start Milestone	0	13-Oct-15	
A1010	Project Management	Level of Effort	5	13-Oct-15	17-Oct-15
A1020	Project Complete	Finish Milestone	0		13-Oct-15
Grading			3	13-Oct-15	15-Oct-15
A1030	Survey and mark out site	Task Dependent	3	13-Oct-15	15-Oct-15
A1040	Site Demarcation	Task Dependent	1	13-Oct-15	13-Oct-15
A1050	Excavation	Resource Dependent	3	13-Oct-15	15-Oct-15
A1060	Grade Site	Task Dependent	2	13-Oct-15	14-Oct-15
Foundation			7	13-Oct-15	21-Oct-15
A1070	Laying of Foundation	Task Dependent	4	13-Oct-15	16-Oct-15
A1080	Slab Plumbing	Task Dependent	7	13-Oct-15	21-Oct-15
A1090	Pour and Float slab concrete	Task Dependent	2	13-Oct-15	14-Oct-15
Structure			30	13-Oct-15	18-Nov-15
Framing			30	13-Oct-15	18-Nov-15
A1100	Exterior Walls	Task Dependent	20	13-Oct-15	05-Nov-15
A1110	Interior Walls	Task Dependent	25	13-Oct-15	12-Nov-15
A1120	Exterior Cladding	Task Dependent	30	13-Oct-15	18-Nov-15
Roof			15	13-Oct-15	30-Oct-15
A1130	Truss Placing	Task Dependent	15	13-Oct-15	30-Oct-15
A1140	Roof Sheeting	Task Dependent	10	13-Oct-15	23-Oct-15
A1150	Place Paper and Shingles	Task Dependent	3	13-Oct-15	15-Oct-15
Interior			10	13-Oct-15	23-Oct-15
Mechanical			3	13-Oct-15	15-Oct-15
A1160	HVAC Ducting	Task Dependent	3	13-Oct-15	15-Oct-15
A1170	Install HVAC Unit	Task Dependent	2	13-Oct-15	14-Oct-15
Electrical			7	13-Oct-15	21-Oct-15
A1180	Breaker Box and Rough Wire	Task Dependent	5	13-Oct-15	17-Oct-15
A1190	Finish Wiring	Task Dependent	7	13-Oct-15	21-Oct-15
Plumbing			10	13-Oct-15	23-Oct-15
A1200	Install Rough plumbing lines	Task Dependent	10	13-Oct-15	23-Oct-15
Decor			10	13-Oct-15	23-Oct-15
A1210	Placing Doors	Task Dependent	8	13-Oct-15	22-Oct-15
A1220	Door Casings and Baseboards	Task Dependent	6	13-Oct-15	20-Oct-15
A1230	Kitchen Cabinets	Task Dependent	10	13-Oct-15	23-Oct-15
Exterior			12	13-Oct-15	27-Oct-15
A1240	Place Window	Task Dependent	12	13-Oct-15	27-Oct-15
A1250	Window Sidings	Task Dependent	5	13-Oct-15	17-Oct-15
Landscaping			5	13-Oct-15	17-Oct-15
A1260	Tree Plantings	Task Dependent	5	13-Oct-15	17-Oct-15

Figure 4-26 *The **Activities** window with all activities*

Exporting the Project

In this section, the created activities will be exported so as to save the project file for future reference. Before exporting, ensure that the **Home Construction** project is opened in Primavera P6.

1. To export a project, choose the **Export** option from the **File** menu; the **Export** wizard with the **Export Format** page is displayed.

2. In this page, ensure that the **Primavera PM/MM - (XER)** radio button is selected and then choose the **Next** button; the **Export Type** page is displayed.

3. In this page, ensure that the **Project** radio button is selected and then choose the **Next** button; the **Projects To Export** page with the Home Construction project is displayed.

4. Choose the **Next** button; the **File Name** page is displayed.

5. In this page, choose the Browse button adjacent to the **File Name** edit box; the **Save File** dialog box is displayed.

6. Browse to *C:/PM6* and then create a sub-folder with the name **c04**. Next, save the file with the name **c04_CONS_Home_Activities_tut01** and then choose the **Save** button.

7. Choose the **Finish** button; the **Primavera P6** message box is displayed with the message that the export was successful.

8. Choose the **OK** button; the file is exported.

Tutorial 2	Relationship and Constraints

In this tutorial, you will establish relationship between activities and then you will assign the constraints to them in the **Home Construction** project.   **(Expected time: 40 min)**

The following steps are required to complete this tutorial:

a. Open the **Activities** window.
b. Establish relationship.
c. Schedule a project.
d. Export the project.

Opening the Activities Window
1. Open Primavera P6 and invoke the **Activities** window by choosing the **Activities** tab from the Directory bar.

Establishing Relationship
1. In the **Activities** window, ensure that the **Activity Details** table is displayed. If not, then choose the **Activity Details** button from the **Activity** toolbar to display it.

2. In the **Activity Details** table, ensure that the **Relationships** tab is available. If it is not available, then right-click on the bar next to the tabs; a shortcut menu is displayed. From the menu, choose the **Customize Activity Details** option; the **Activity Details** dialog box is displayed. Select the **Relationships** option from the **Available Tabs** area and then choose the **Add to list** button; the option is added into the **Display Tabs** area. Choose the **OK** button; the tab is displayed in the **Activity Details** table.

3. Choose the **Relationships** tab; the **Predecessors** and **Successors** area is displayed.

4. In the **Activities** window, ensure that the **Project Start** activity is selected.

5. Now, in the **Successors** area, choose the **Assign** button; the **Assign Successors** dialog box is displayed, as shown in Figure 4-27.

6. In this dialog box, select the **Project Management** option from the **Activity Name** column and choose the **Assign** button.

7. Choose the **Close** button; the dialog box is closed and an activity is added under the **Successors** area.

8. Click on the cells under the **Relations** column; a drop-down list is displayed.

9. Select the **SS** option from the **Relations** drop-down list; the activity is assigned the selected relationship, as shown in Figure 4-28.

Figure 4-27   The Assign Successors dialog box

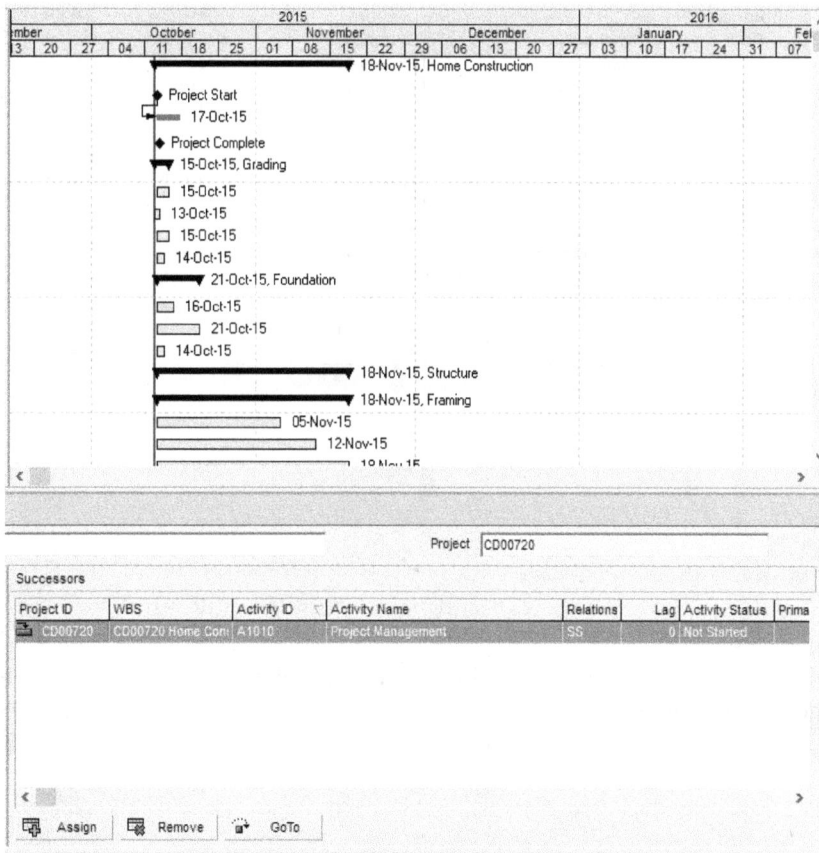

*Figure 4-28   The **Successors** area displaying activity with the Gantt chart*

Note

*Ensure that the **Relationship Lines** option is selected, to view the relationship lines in the Gantt Chart*

If you select the **Project Management** activity under the **Home Construction** WBS, the **Project Start** activity is displayed in the **Predecessors** area, as shown in Figure 4-29. This indicates that the **Project Start** activity is the predecessor of **Project Management** activity.

*Figure 4-29 The **Project Start** activity shown in the **Predecessors** area*

10. Repeat the procedure followed in step 1 through step 9 and assign the successor relationship to the activity, refer to Table 4-1.

Table 4-1 *The predecessor and successor relationships*

Predecessor ID	Predecessor Name	Successor ID	Successor Name	Relationship Type	Lag
EC00720 New Home Construction					
A1000	Project Start	A1010	Project Management	SS	0
A1000	Project Start	A1030	Survey and Mark Out Site	SS	0
A1010	Project Management	A1020	Project Complete	FF	0
EC00720.1 Grading					
A1030	Survey and Mark Out Site	A1040	Site Demarcation	FS	0
A1040	Site Demarcation	A1050	Excavation	FS	1
A1050	Excavation	A1060	Grade Site	FS	0
EC00720.2 Foundation					
A1070	Laying of Foundation	A1080	Slab Plumbing	FS	0
A1080	Slab Plumbing	A1090	Pour and Float Slab Concrete	FS	0
A1090	Pour and Float Slab Concrete	A1100	Exterior Walls	FS	7
EC00720.3.1 Framing					
A1100	Exterior Walls	A1110	Interior Walls	FS	0
A1100	Exterior Walls	A1130	Truss placing	FS	0
A1100	Exterior Walls	A1180	Breaker Box and Rough Wiring	FS	0
A1100	Exterior Walls	A1200	Install rough plumbing lines	FS	0
A1100	Exterior Walls	A1120	Exterior Cladding	FS	0
A1120	Exterior Cladding	A1240	Place Windows	FS	0
EC00720.3.2 Roof					
A1130	Truss Placing	A1140	Roof Sheeting	FS	0
A1140	Roof Sheeting	A1150	Place Paper and Shingles	FS	0
A1150	Place Paper and Shingles	A1160	HVAC Ducting	FS	0
EC00720.3.3.1 Mechanical					
A1160	HVAC Ducting	A1170	Install HVAC Unit	SS	0
A1170	Install HVAC Unit	A1020	Project Complete	FF	0
EC00720.3.3.2 Electrical					
A1180	Breaker Box & Rough Wiring	A1190	Finish Wiring	FS	0
A1190	Finish Wiring	A1020	Project Complete	FF	0
EC00720.3.3.3 Plumbing					
A1200	Install Rough Plumbing Lines	A1210	Placing Doors	FS	0
EC00720.3.3.5 Decor					
A1210	Placing Doors	A1220	Door Casings and Baseboards	FS	0
A1230	Kitchen cabinets	A1020	Project Complete	FF	0
EC00720.3.4 Exterior					
A1240	Place Windows	A1250	Window Siding	FS	0
A1250	Window Siding	A1260	Tree Plantings	FS	0
EC00720.4 Landscaping					
A1260	Tree Plantings	A1020	Project Complete	FF	0

Figure 4-30 shows the **Activities** window and Gantt chart after applying the relationships.

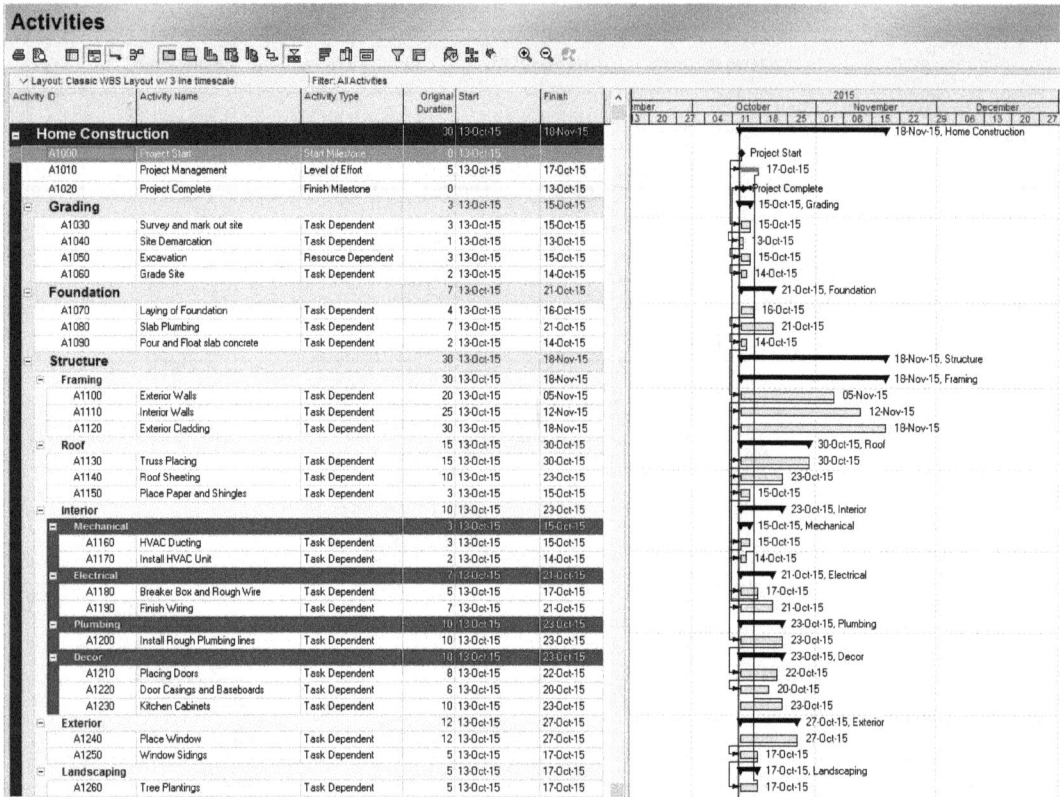

Figure 4-30 The Activities window with the Gantt chart

Scheduling a Project

1. Choose the **Schedule** option from the **Tools** menu; the **Schedule** dialog box is displayed.

2. In the dialog box, choose the Browse button adjacent to the **Current Data Date** edit box and change the date to 13-Oct-15 and choose the **Select** button.

3. Choose the **Schedule** button; the project is updated with the scheduled date, as shown in Figure 4-31.

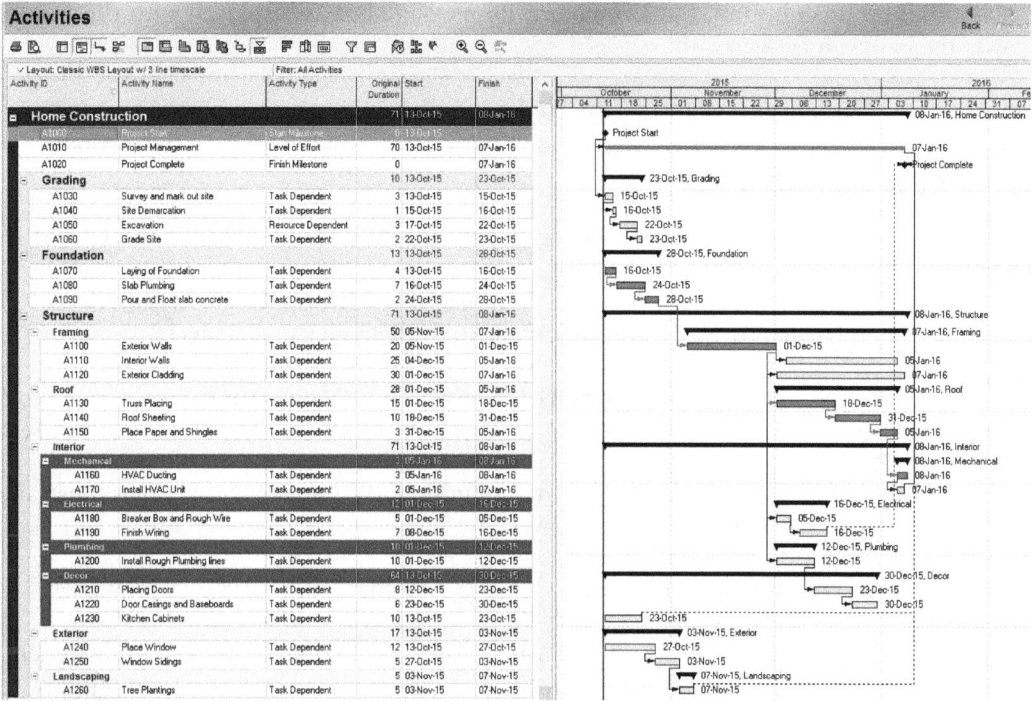

Figure 4-31 *The* **Activities** *window updated with the date*

Exporting the Project

You need to export projects to save the project file for future reference. Ensure that the Home Construction project is opened in Primavera P6.

1. To export a project, choose the **Export** option from the **File** menu; the **Export** wizard with the **Export Format** page is displayed.

2. In this page, ensure that the **Primavera PM/MM - (XER)** radio button is selected and then choose the **Next** button; the **Export Type** page is displayed.

3. In this page, ensure that the **Project** radio button is selected and then choose the **Next** button; the **Projects To Export** page with the **Home Construction** project is displayed.

4. Choose the **Next** button; the **File Name** page is displayed.

5. In this page, choose the Browse button adjacent to the **File Name** edit box; the **Save File** dialog box is displayed.

6. Browse to *C:/PM6/c04* and save the file with the name **c04_CONS_Home_tut02** and then choose the **Save** button.

7. Choose the **Finish** button; the **Primavera P6** message box is displayed with the message that the export was successful.

8. Choose the **OK** button; the file is exported.

Self-Evaluation Test

Answer the following questions and then compare them to those given at the end of this chapter:

1. The _____ table is used to add the details to the activities in a project.

2. The _____ activity type is used to mark the beginning of a phase or communicate project deliverable.

3. On choosing the _____ tab, actual start and finish dates, free float, total float, constraints, and duration of the project will be displayed.

4. The **Columns** button is available in the _____ toolbar.

5. The _____, _____, _____, and _____ relationship types are assigned to activities from the **Relationships** tab.

6. You can build a project schedule and schedule it in Primavera P6. (T/F)

7. Activities can help in planning a project. (T/F)

8. Any error in the relationship of activity is highlighted in red on the Gantt chart. (T/F)

9. The critical path in activities will be displayed only when the activities are linked incorrectly. (T/F)

10. You can establish relationships between activities in the same project. (T/F)

Review Questions

Answer the following questions:

1. Which of the following software can be used to generate the project schedule?

    (a) Microsoft Excel   (b) Microsoft Project
    (c) Primavera   (d) Microsoft Office

2. Which of the following type of relationships is assigned in an activity?

    (a) FA   (b) FD
    (c) AS   (d) SS

3. Which of the following is a type of activity?

    (a) Planned Start   (b) Level of Effort
    (c) Task Milestone   (d) Planned Finish

4. Which of the following tabs is used to establish a relationship?

    (a) Status   (b) Resources
    (b) Relationship   (d) Steps

5. The **Activity Network** is used to visualize the flow of logic as you link activities. (T/F)

6. You can schedule the relationship calendar from the **Schedule Options** dialog box. (T/F)

7. In Primavera P6, constraints cannot be assigned to activities. (T/F)

8. You cannot update a relationship once it has been established. (T/F)

9. You can edit the activity name and ID after they have been assigned. (T/F)

EXERCISE

Exercise 1

In this exercise, you will add activities to the project created in Exercise 1 of chapter 3 of the **Hospital Building** project.   **(Expected time: 60 min)**

     Hints
1. Start Primavera P6 and open the **Projects** window.
2. Open the **Hospital Building** project.
3. Open the **Activities** window.
4. Create Activities, as shown in Figures 4-32 and 4-33.
5. Export a project.
6. Save the file with the name *c04_CONS_HOSP_ex01*.

Note that, the activities of Upper Basement are the same as that of Lower Basement and the activities of First Floor, Second Floor, Third Floor, and LMR/Mumty activities are the same as that of Ground Floor.

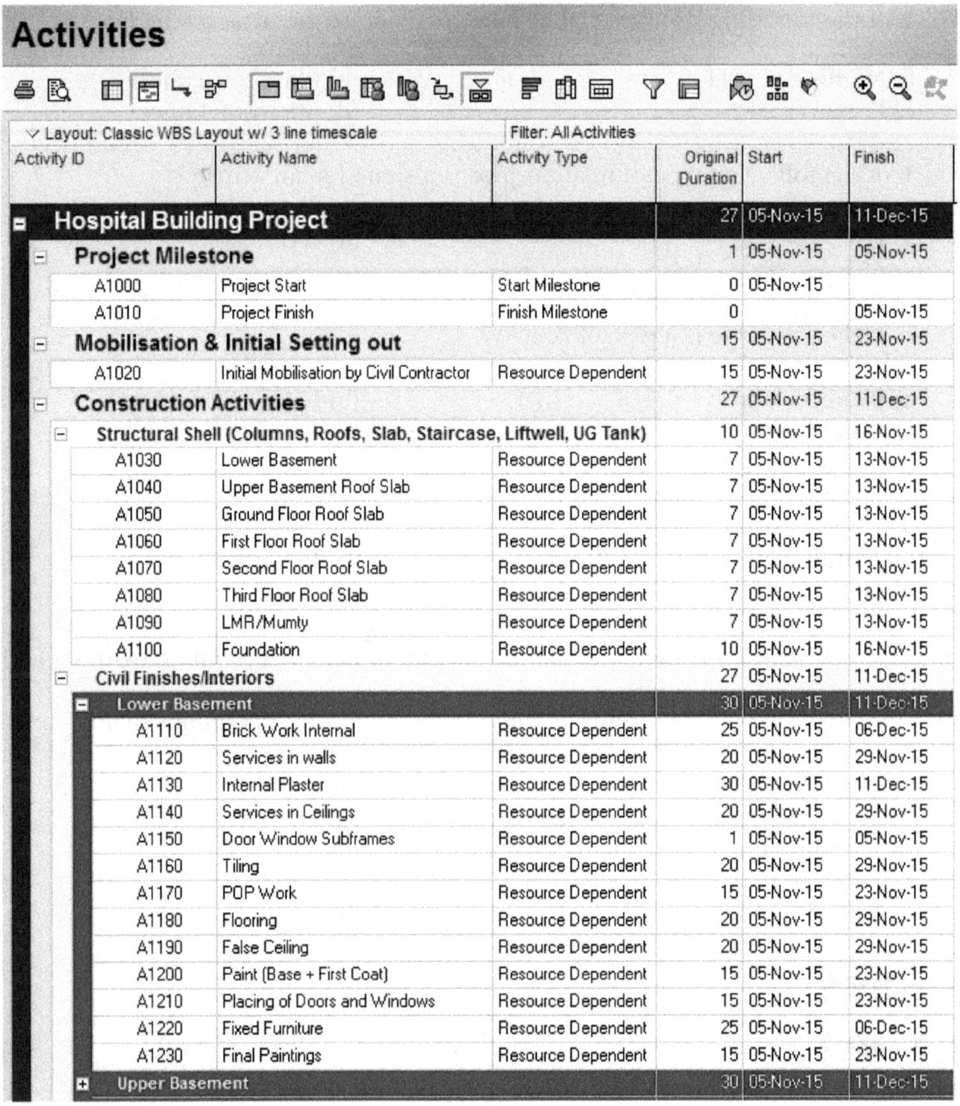

Activities

Activity ID	Activity Name	Activity Type	Original Duration	Start	Finish
Hospital Building Project			27	05-Nov-15	11-Dec-15
Project Milestone			1	05-Nov-15	05-Nov-15
A1000	Project Start	Start Milestone	0	05-Nov-15	
A1010	Project Finish	Finish Milestone	0		05-Nov-15
Mobilisation & Initial Setting out			15	05-Nov-15	23-Nov-15
A1020	Initial Mobilisation by Civil Contractor	Resource Dependent	15	05-Nov-15	23-Nov-15
Construction Activities			27	05-Nov-15	11-Dec-15
Structural Shell (Columns, Roofs, Slab, Staircase, Liftwell, UG Tank)			10	05-Nov-15	16-Nov-15
A1030	Lower Basement	Resource Dependent	7	05-Nov-15	13-Nov-15
A1040	Upper Basement Roof Slab	Resource Dependent	7	05-Nov-15	13-Nov-15
A1050	Ground Floor Roof Slab	Resource Dependent	7	05-Nov-15	13-Nov-15
A1060	First Floor Roof Slab	Resource Dependent	7	05-Nov-15	13-Nov-15
A1070	Second Floor Roof Slab	Resource Dependent	7	05-Nov-15	13-Nov-15
A1080	Third Floor Roof Slab	Resource Dependent	7	05-Nov-15	13-Nov-15
A1090	LMR/Mumty	Resource Dependent	7	05-Nov-15	13-Nov-15
A1100	Foundation	Resource Dependent	10	05-Nov-15	16-Nov-15
Civil Finishes/Interiors			27	05-Nov-15	11-Dec-15
Lower Basement			30	05-Nov-15	11-Dec-15
A1110	Brick Work Internal	Resource Dependent	25	05-Nov-15	06-Dec-15
A1120	Services in walls	Resource Dependent	20	05-Nov-15	29-Nov-15
A1130	Internal Plaster	Resource Dependent	30	05-Nov-15	11-Dec-15
A1140	Services in Ceilings	Resource Dependent	20	05-Nov-15	29-Nov-15
A1150	Door Window Subframes	Resource Dependent	1	05-Nov-15	05-Nov-15
A1160	Tiling	Resource Dependent	20	05-Nov-15	29-Nov-15
A1170	POP Work	Resource Dependent	15	05-Nov-15	23-Nov-15
A1180	Flooring	Resource Dependent	20	05-Nov-15	29-Nov-15
A1190	False Ceiling	Resource Dependent	20	05-Nov-15	29-Nov-15
A1200	Paint (Base + First Coat)	Resource Dependent	15	05-Nov-15	23-Nov-15
A1210	Placing of Doors and Windows	Resource Dependent	15	05-Nov-15	23-Nov-15
A1220	Fixed Furniture	Resource Dependent	25	05-Nov-15	06-Dec-15
A1230	Final Paintings	Resource Dependent	15	05-Nov-15	23-Nov-15
Upper Basement			30	05-Nov-15	11-Dec-15

*Figure 4-32 The partial view of the **Activities** window with the activities till Upper Basement*

⊟	**Ground Floor**			30	05-Nov-15	11-Dec-15
	A1370	Brickwork Internal/External	Resource Dependent	25	05-Nov-15	06-Dec-15
	A1380	Services in walls	Resource Dependent	20	05-Nov-15	29-Nov-15
	A1390	Plaster Internal/External	Resource Dependent	30	05-Nov-15	11-Dec-15
	A1400	Services in Ceiling	Resource Dependent	20	05-Nov-15	29-Nov-15
	A1410	Door Window Subframes	Resource Dependent	1	05-Nov-15	05-Nov-15
	A1420	Tiling	Resource Dependent	20	05-Nov-15	29-Nov-15
	A1430	POP Work	Resource Dependent	15	05-Nov-15	23-Nov-15
	A1440	Flooring	Resource Dependent	20	05-Nov-15	29-Nov-15
	A1450	False Ceiling	Resource Dependent	20	05-Nov-15	29-Nov-15
	A1460	Paint (Base + First Coat)/External Paint	Resource Dependent	15	05-Nov-15	23-Nov-15
	A1470	Placing of Doors and Windows	Resource Dependent	15	05-Nov-15	23-Nov-15
	A1480	Fixed Furniture	Resource Dependent	25	05-Nov-15	06-Dec-15
	A1490	Final Paintings	Resource Dependent	15	05-Nov-15	23-Nov-15
⊞	**First Floor**			30	05-Nov-15	11-Dec-15
⊞	**Second Floor**			30	05-Nov-15	11-Dec-15
⊞	**Third Floor**			30	05-Nov-15	11-Dec-15
⊞	**LMR/Mumty**			30	05-Nov-15	11-Dec-15
⊟	**Other Services**			25	05-Nov-15	06-Dec-15
	A1990	Electrical	Resource Dependent	25	05-Nov-15	06-Dec-15
	A2000	Plumbing	Resource Dependent	25	05-Nov-15	06-Dec-15
	A2010	HVAC	Resource Dependent	25	05-Nov-15	06-Dec-15
⊟	**Lifts/Elevators**			3	05-Nov-15	08-Nov-15
	A2020	Delivery of Lifts	Resource Dependent	1	05-Nov-15	05-Nov-15
	A2030	Installations and Commisioning of Lifts	Resource Dependent	3	05-Nov-15	08-Nov-15
⊟	**External Developments**			20	05-Nov-15	29-Nov-15
	A2040	Landscape and Horticulture	Resource Dependent	20	05-Nov-15	29-Nov-15
	A2050	Roads and external gates	Resource Dependent	15	05-Nov-15	23-Nov-15
	A2060	External Lighting and Signages	Resource Dependent	10	05-Nov-15	16-Nov-15
⊟	**Handing Over**			10	05-Nov-15	16-Nov-15
	A2070	Loose Furniture	Resource Dependent	10	05-Nov-15	16-Nov-15
	A2080	Ready for user trial	Resource Dependent	10	05-Nov-15	16-Nov-15

*Figure 4-33 The partial view of the **Activities** window showing the Ground Floor onward activities*

Note that under the **LMR/Mumty** head the **Tiling**, **Fixed Furniture**, and **False Ceiling** activities will not be added.

Exercise 2

In this exercise, you will assign the relationship between the activities created in Exercise 1 of this chapter. Refer to Table 4-2 for the relationship to be assigned between different activities.

(Expected time: 60 min)

Table 4-2 *The relationship between activities.*

Activity ID	Activity Name	Successor ID	Successor Name	Relationship Type	Lag
CD00770 Hospital Building Project					
A1000	Project Start	A1020	Initial Mobilisation by Civil Contractor	FS	0
		A1990	Electrical	FS	97
A1020	Initial Mobilisation by Civil Contractor	A1100	Foundation	FS	0
Construction Activities					
A1030	Lower Basement Roof Slab	A1040	Upper Basement Roof Slab	FS	0
		A1110	Brick Work Internal	FS	14
A1040	Upper Basement Roof Slab	A1050	Ground Floor Roof Slab	FS	0
		A1110	Brick Work Internal	FS	14
A1050	Ground Floor Roof Slab	A1060	First Floor Roof Slab	FS	0
		A1370	Brick Work Internal/External (G.F)	FS	14
A1060	First Floor Roof Slab	A1070	Second Floor Roof Slab	FS	0
		A1500	Brick Work Internal/External (F.F)	FS	63
A1070	Second Floor Roof Slab	A1080	Third Floor Roof Slab	FS	0
		A1630	Brick Work Internal/External (S.F)	FS	14
A1080	Third Floor Roof Slab	A1090	LMR/Mumty	FF	5
		A1760	Brick Work Internal/External (T.F)	FS	14
A1090	LMR/Mumty	A1890	Brick Work Internal/External (LMR/Mumty)	FS	0
		A2020	Delivery of Lifts	FS	43
A1100	Foundation	A1030	Lower Basement Roof Slab	FS	0
Civil Finishes / Interiors					
Lower Basement					
A1110	Brick Work Internal	A1120	Service in walls	FF	23
		A1130	Internal Plaster	FF	7
		A1240	Brick Work Internal (Upper Basement)	FS	0
A1120	Service in walls	A1130	Internal Plaster	FF	7
		A1140	Services in Ceiling	SS	3
		A1160	Tiling	FF	7
A1130	Internal Plaster	A1150	Door Window Subframes	FF	0
		A1170	POP Work	FF	7
A1140	Services in Ceiling	A1180	Flooring	FF	13
		A1190	False Ceiling	FF	10
A1150	Door Window Subframes	A1160	Tiling	FF	0
		A1170	POP Work	FF	7
A1160	Tiling	A1170	POP Work	FF	7
A1170	POP Work	A1180	Flooring	FF	10
		A1200	Paint (Base + First Coat)	FF	5
A1180	Flooring	A1190	FalseCeiling	FF	10
A1190	FalseCeiling	A1200	Paint (Base + First Coat)	FF	5
A1200	Paint (Base + First Coat)	A1210	Placing of Doors and Windows	FF	7
A1210	Placing of Doors and Windows	A1220	Fixed Furniture	FF	7
A1220	Fixed Furniture	A1230	Final Paintings	FF	7

Upper Basement					
A1240	Brick Work Internal	A1250	Service in walls	FF	12
		A1260	Internal Plaster	FF	7
A1250	Service in walls	A1260	Internal Plaster	FF	7
		A1270	Services in Ceiling	SS	3
		A1290	Tiling	FF	7
A1260	Internal Plaster	A1280	Door Window Subframes	FF	0
		A1300	POP Work	FF	7
A1270	Services in Ceiling	A1310	Flooring	FF	13
		A1320	False Ceiling	FF	10
A1280	Door Window Subframes	A1290	Tiling	FF	0
		A1300	POP Work	FF	7
A1290	Tiling	A1300	POP Work	FF	7
A1300	POP Work	A1310	Flooring	FF	10
		A1330	Paint (Base + First Coat)	FF	5
A1310	Flooring	A1320	FalseCeiling	FF	10
A1320	FalseCeiling	A1330	Paint (Base + First Coat)	FF	5
A1330	Paint (Base + First Coat)	A1340	Placing of Doors and Windows	FF	3
A1340	Placing of Doors and Windows	A1350	Fixed Furniture	FF	7
A1350	Fixed Furniture	A1360	Final Paintings	FF	7
A1360	Final Paintings	A2070	Loose Furniture	FF	3
Ground Floor					
A1370	Brickwork Internal/External	A1380	Service in walls	FF	12
		A1390	Plaster Internal / External	FF	7
A1380	Services in walls	A1390	Plaster Internal / External	FF	7
		A1400	Services in Ceiling	SS	3
		A1420	Tiling	FF	7
A1390	Plaster Internal / External	A1410	Door Window Subframes	FF	45
		A1430	POP Work	FF	7
A1400	Services in Ceiling	A1440	Flooring	FS	13
		A1450	False Ceiling	FF	10
A1410	Door Window Subframes	A1420	Tiling	FF	0
		A1430	POP Work	FF	7
A1420	Tiling	A1430	POP Work	FF	7
A1430	POP Work	A1440	Flooring	FF	10
		A1460	Paint (Base + First Coat)/External Paint	FS	5
A1440	Flooring	A1450	False Ceiling	FF	10
A1450	False Ceiling	A1460	Paint (Base + First Coat)/External Paint	FF	5
A1460	Paint (Base + First Coat)/External Paint	A1470	Placing of Doors and Windows	FF	3
A1470	Placing of Doors and Windows	A1480	Fixed Furniture	FF	7
A1480	Fixed Furniture	A1490	Final Paintings	FF	7
A1490	Final Paintings	A2070	Loose Furniture	FF	3

First Floor					
A1500	Brickwork Internal/External	A1510	Services in walls	FF	33
		A1520	Plaster Internal / External	FF	7
A1510	Service in walls	A1520	Plaster Internal / External	FF	7
		A1530	Services in Ceiling	SS	3
		A1550	Tiling	FF	7
A1520	Plaster Internal / External	A1540	Door Window Subframes	FF	0
		A1560	POP Work	FF	7
A1530	Services in Ceiling	A1570	Flooring	FF	44
		A1580	False Ceiling	FF	10
A1540	Door Window Subframes	A1550	Tiling	FF	0
		A1560	POP Work	FF	7
A1550	Tiling	A1560	POP Work	FF	7
A1560	POP Work	A1570	Flooring	FF	10
		A1590	Paint (Base + First Coat)/External Paint	FF	5
A1570	Flooring	A1580	False Ceiling	FF	10
A1580	False Ceiling	A1590	Paint (Base + First Coat)/External Paint	FF	5
A1590	Paint (Base + First Coat)/External Paint	A1600	Placing of Doors and Windows	FF	3
A1600	Placing of Doors and Windows	A1610	Fixed Furniture	FF	7
A1610	Fixed Furniture	A1620	Final Paintings	FF	7
A1620	Final Paintings	A2100	Loose Furniture	FF	3
Second Floor					
A1630	Brickwork Internal/External	A1640	Service in walls	FF	22
		A1650	Plaster Internal / External	FF	7
A1640	Service in walls	A1650	Plaster Internal / External	FF	7
		A1660	Services in Ceiling	SS	3
		A1680	Tiling	FF	7
A1650	Plaster Internal / External	A1670	Door Window Subframes	FF	0
		A1690	POP Work	FF	7
A1660	Services in Ceiling	A1700	Flooring	FF	10
		A1710	False Ceiling	FF	10
A1670	Door Window Subframes	A1680	Tiling	FF	0
		A1690	POP Work	FF	7
A1680	Tiling	A1690	POP Work	FF	7
A1690	POP Work	A1700	Flooring	FF	10
		A1720	Paint (Base + First Coat)/External Paint	FF	5
A1700	Flooring	A1710	False Ceiling	FF	10
A1710	False Ceiling	A1720	Paint (Base + First Coat)/External Paint	FF	5
A1720	Paint (Base + First Coat)/External Paint	A1730	Placing of Doors and Windows	FF	3
A1730	Placing of Doors and Windows	A1740	Fixed Furniture	FF	7
A1740	Fixed Furniture	A1750	Final Paintings	FF	7
A1750	Final Paintings	A2100	Loose Furniture	FF	3

Third Floor					
A1760	Brickwork Internal/External	A1770	Service in walls	FF	27
		A1780	Plaster Internal / External	FF	7
A1770	Service in walls	A1780	Plaster Internal / External	FF	7
		A1790	Services in Ceiling	SS	3
		A1810	Tiling	FF	7
A1780	Plaster Internal / External	A1800	Door Window Subframes	FF	0
		A1820	POP Work	FF	7
A1790	Services in Ceiling	A1830	Flooring	FF	10
		A1840	False Ceiling	FF	10
A1800	Door Window Subframes	A1810	Tiling	FF	0
		A1820	POP Work	FF	7
A1810	Tiling	A1820	POP Work	FF	7
A1820	POP Work	A1830	Flooring	FF	10
		A1850	Paint (Base + First Coat)/External Paint	FF	5
A1830	Flooring	A1840	False Ceiling	FF	10
A1840	False Ceiling	A1850	Paint (Base + First Coat)/External Paint	FF	5
A1850	Paint (Base + First Coat)/External Paint	A1860	Placing of Doors and Windows	FF	3
A1860	Placing of Doors and Windows	A1870	Fixed Furniture	FF	7
A1870	Fixed Furniture	A1880	Final Paintings	FF	7
A1880	Final Paintings	A2070	Loose Furniture	FF	3
LMR/Mumty					
A1890	Brickwork Internal/External	A1900	Service in walls	FF	17
		A1910	Plaster Internal / External	FF	7
A1900	Service in walls	A1910	Plaster Internal / External	FF	7
		A1920	Services in Ceiling	SS	3
A1910	Plaster Internal / External	A1930	Door Window Subframes	FF	0
		A1940	POP Work	FF	7
A1920	Services in Ceiling	A1950	Flooring	FF	10
A1930	Door Window Subframes	A1940	POP Work	FF	7
A1950	POP Work	A1950	Flooring	FF	10
		A1960	Paint (Base + First Coat)/External Paint	FF	5
A1950	Flooring	A1960	Paint (Base + First Coat)/External Paint	FF	7
A1960	Paint (Base + First Coat)/External Paint	A1970	Placing of Doors and Windows	FF	3
A1970	Placing of Doors and Windows	A1980	Final Paintings	FF	7
Other Services					
A1990	Electrical	A2000	Plumbing	FF	12
A2000	Plumbing	A2010	HVAC	FF	12
A2010	HVAC	A1110	Brick Work Internal	SS	7
		A1450	False Ceiling	FF	7
Lifts & Elevators					
A2020	Delivery of Lifts	A2030	Installation and Commisioning of Lifts	FF	3
External Developments					
A2040	Landscape and Horticulture	A2050	Roads and external gates	FS	0
A2050	Roads and external gates	A2060	External Lighting and Signages	FS	0
Handing Over					
A2080	Ready for user trial	A1010	Project Finish	FF	0

Answers to Self-Evaluation Test

1. Activity Details, **2.** Start Milestone, **3.** Status, **4.** Activity, **5.** FF, FS, SS, SF, **6.** T, **7.** T, **8.** T, **9.** T, **10.** T

Chapter 5

Defining Resources and Roles

Learning Objectives

After completing this chapter, you will be able to:
- *Understand the resources*
- *Manage resources*
- *Understand Resources Layout*
- *Assign roles to resources*
- *Assign resources to project*

INTRODUCTION

In the previous chapter, you learned about the features and details of the activities layout. You also learned to add activities under a WBS. Moreover, you will assign relationships in terms of predecessor and successor between activities.

In this chapter, you will learn to add resources to activities and to assign roles to the resources to complete activities. Moreover, you will learn the tools used for managing resources and roles across a project or enterprise.

RESOURCES

Resources are the entities that are required to perform an activity. These resources such as manpower, equipment, and materials are very important to carry out activities as their lack can affect the project progess. To track the progress of project, scheduling should be done. You can schedule projects by assigning or even without assigning resources to activities as only relationship between activities is needed to schedule the project.

Resources are generally reused between activities and projects. In Project Management, you can create your own resources that can be used in the entire project or activities. Additionally, you can distinguish between labor, material, and nonlabor resources. Labor and nonlabor resources are always time-based, whereas material resources are consumable items and are measured in units. You can create resource calendars, assign resources to different activities, and also define the roles, contact information, and time-varying prices of the resources.

Resources are categorized as: labor, nonlabor, and material.

Labor Resources: Labor resources such as managers, operators, subcontractors, and so on, are the people who manage and execute the activities.

Non-Labor Resources: The non labor resources such as machinery, scaffold, or other equipment, are those resources which are used to perform activities. These are measured in time units.

Material Resources: This resource type includes the items that are consumed in performing an activity or work such as piping, wires, tubing that is being installed. The material resources are measured in units such as cubic yards or in cubic meters.

Resources are time-based and generally extended across multiple activities and/or projects, while expenses are one-time expenditures for nonreusable items required by activities.

Managing Resources

In Primavera P6, you can assign resources to the activities of the project by using the **Resources** window. This window enables you to manage the resource hierarchy and individual resource information. You can assign resources to activities and roles in any project. If resource security is enabled, you can view the resources you have access to.

To display the **Resources** window, choose the **Resources** tab from the Directory bar. Alternatively, you can choose the **Resources** option from the **Enterprise** menu in the menubar. The **Resources** window enables you to view and add the resources to complete projects. The **Resources** window displays the **Resource** layout and the **Resources Details** table, refer to Figure 5-1.

In the **Resources** window, you can view the resource hierarchy as a table or as a chart. In addition, you can restructure the hierarchy by repositioning the resources at different levels according to the work performed and other criteria.

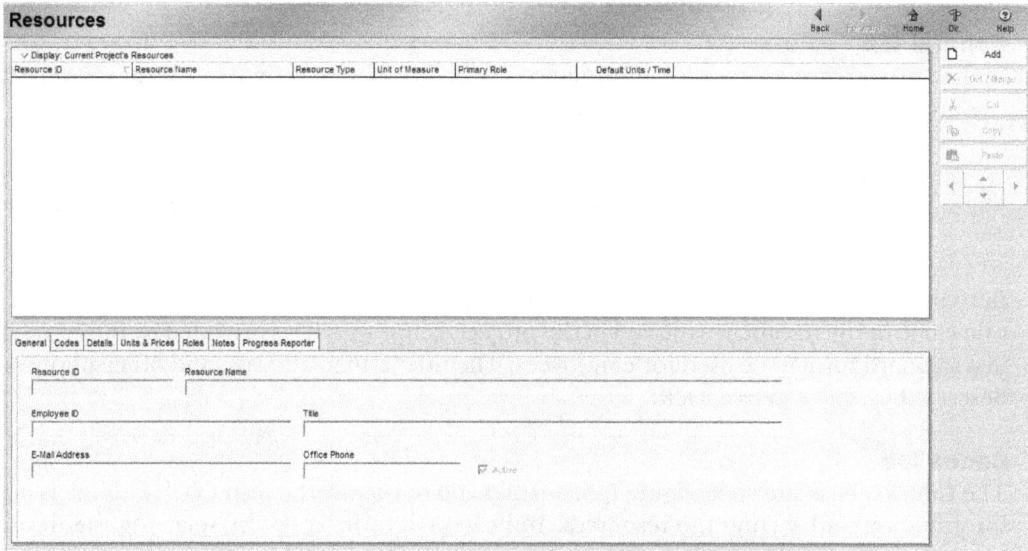

*Figure 5-1 The **Resources** window with the **Resources Details** table*

Creating Resources Layout

In the **Resources** window, you need to set the display of the window by selecting the required layout. The **Current Project's Resources** option is selected by default in the **Display** drop-down. If there are no resources in the opened project, the **Resources** window will not display any resource. Note that, if you add resources to this window, the **Information** message box will be displayed and you will be prompted to change the display to **All Resources**. To change the resource display, choose the **All Resources** option from the **Filter By** cascading menu of the **Display** drop-down; all resources will be displayed. In the displayed resources, you can customize the columns of the resources. To customize the columns, choose the **Customize** option from the **Columns** cascading menu of the **Display** drop-down; the **Columns** dialog box will be displayed. In this dialog box, move the required options from the **Available Options** area to the **Selected Options** area by using the **Add to list** button. Next, choose the **OK** button; the dialog box will be closed.

The **Resources** window displays the **Resource Details** table with various tabs by default. In these tabs, you can provide general information, codes, and details of the resources and so on. In case the **Resources Details** table is not displayed by default, choose the **Details** option from the **Display** drop-down; the **Resources Details** table will be displayed, as shown in Figure 5-2. The tabs of the table are described in detail next.

Figure 5-2 *The **Resources Details** table*

General Tab

In this tab, you can view and edit information about the selected resources such as resource ID, resource name, employee ID, e-mail address, and so on.

> **Note**
> *The edit boxes of this tab will be enabled only when a resource is selected in the **Resources** window.*

Resource ID is a global unique identifier for a resource. Resource name is the descriptive title name of the resources assigned in the project. Employee ID is provided by the company in a standard form to be used for employees. The title, e-mail address, and office phone are described by the resource itself.

Codes Tab

The **Codes** tab, as shown in Figure 5-3, enables you to classify the resources. This tab is used for grouping and sorting the resources. You can assign the codes by selecting the desired resource from the **Resources** window and then choose the **Assign** button; the **Assign Resource Codes** dialog box will be displayed. In this dialog box, choose the code from the **Resource Code** column and then choose the **Assign** button; the dialog box will be closed and the codes will be assigned to the resources.

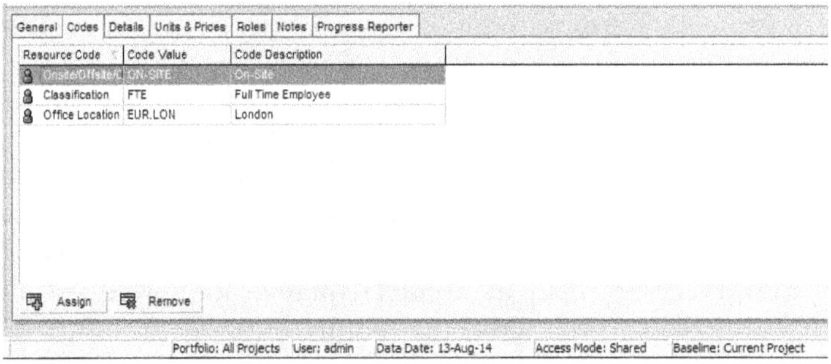

Figure 5-3 *Partial view of the **Codes** tab in the **Resources** Details table*

Details Tab

The **Details** tab allows you to enter the detailed description of the resources. This tab consists of the **Resource Type**, **Currency and Overtime**, and **Profile** areas.

In the **Resource Type** area, you can assign the resource type such as Labor, Nonlabor, and Material to the required resources. The **Currency and Overtime** area helps you to enter the currency rate at which the resource is assigned. In the **Overtime Factor** edit box, you can enter factor of overtime between 1 and 10. The overtime factor is used to calculate how much the labor resource will cost if they work more than the working hours. It is the factor which is to be multiplied with the standard price of the resources for calculating the overtime price **(standard price* overtime factor= overtime price)**. To enter the over time, select the **Overtime Allowed** check box and then enter the factor of overtime in the **Overtime Factor** edit box.

In the **Profile** area, you can assign a calendar to the resources or can edit existing calendar. In this area, you can enter the default unit/time value in the **Default Units/Time** edit box. If you select the **Auto Compute Actuals** check box, the actual quantity of work performed by the resources will be computed automatically. Clear this check box if you are using the Progress Reporter to update actuals. You can select the **Calculate costs from units** check box to calculate the cost depending upon the units of resources assigned. Alternatively, you need to recalculate the cost incurred on resources whenever any quantity is updated.

Units & Prices Tab

The **Units & Prices** tab, as shown in Figure 5-4, is used to define the availability of the resources and their standard price and also to set the calendar for different shifts. To set the calendar for the resource shift, choose the Browse button in the **Shift Calendar** field; the **Select Shift** dialog box will be displayed. In this dialog box, select the shift type from the **Shift Name** column and then choose the **Select** button; the shift will be assigned and the dialog box will be closed. The **Shift** spinner will allow you to enter the number of shifts according to the calendar assigned in the **Shift Calendar** field. The **Effective Date** column determines the date on and after which a resource's maximum units/time and standard rate will be applied. In the **Max Units/Time** column, enter the number of resources working per hour/ per day/per week/ per month. If the resource works for more hours than the assigned time, then it will be considered as overtime. The **Price/Unit** column allows to assign the daily, hourly, weekly, or yearly wages of a resource.

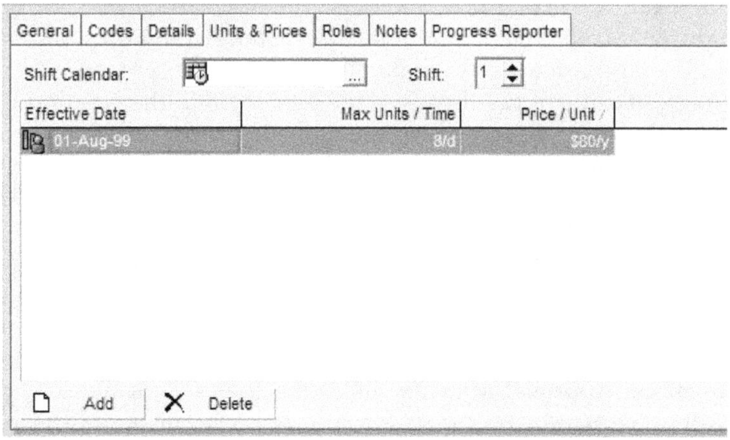

Figure 5-4 The Resources Details table with the Units & Prices tab

Roles Tab

In the **Roles** tab, refer to Figure 5-5, you can set different types of roles for the resource with one role set as the primary (default) role. In addition, you can assign proficiency to a resource for a given role. To assign the proficiency, select the desired option from the **Proficiency** drop-down list. In this drop-down list, the skill level ranges from **1 - Master** to **5 - Inexperienced**. Select the desired option for the resource with the right level of skills for a given task.

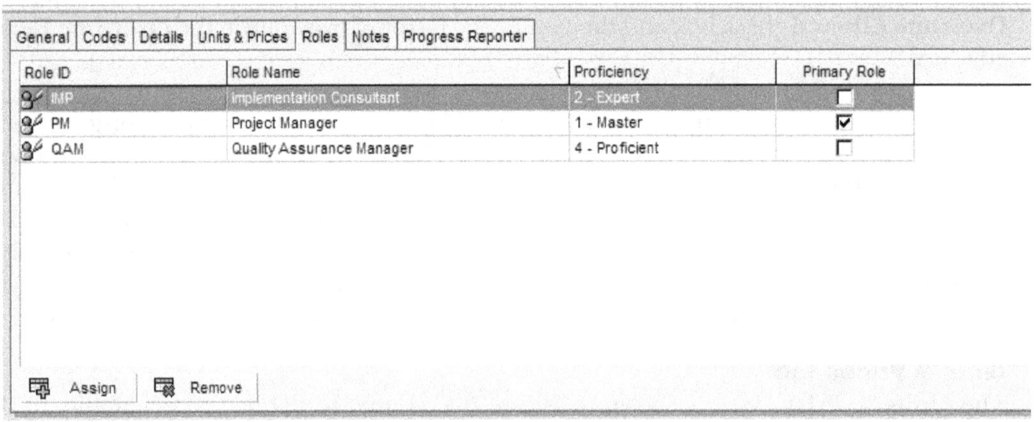

Figure 5-5 The Roles tab in the Resources Details table

Notes Tab
The **Notes** tab will allow you to add notes related to the resources using the writing tools.

Progress Reporter Tab
The **Progress Reporter** tab is divided into the **Login** and **Timesheet** areas. In the **Login** area, you need to specify the user login name by which a user will login into the progress reporter. To select the user login, choose the Browse button from the **User Login** field; the **Select User Login Name** dialog box will be displayed. In this dialog box, select the required user name by which you want to login and then choose the **Select** button; the user name will be assigned and the dialog box will be closed. You can choose the **Edit User** button from the **Login** area to edit the details of the user. Choose the **Edit User** button; the **Users** dialog box will be displayed. In this dialog box, select the user whose filled entries are to be edited. Now, you can edit the details of the user login in the **General**, **Contact**, **Global Access**, **Project Access**, and **Module Access** tabs. You can add your own user by choosing the **Add** button from the Command bar of the **Users** dialog box and fill the necessary details of the newly created user. Choose the **Close** button; the dialog box will be closed.

In the **Timesheet** area, select the **Uses timesheets** check box to indicate that the selected resource uses the timesheet to report workhours in P6 progress reporter. In the **Timesheet Approval Manager** field, you need to assign the name of the user who has the global access privilege to approve timesheets. To assign the manager name, choose the Browse button from the **Timesheet Approval Manager** field; the **Select Timesheet Approval Manager** dialog box will be displayed. In this dialog box, select the name of the user who has the access to approve timesheet.

Adding Resources
To assign resources to the activities, you are required to create resources. To create resources, choose the **Add** button from the Command bar; the **New Resource Wizard** with the **Resource**

ID and Name page will be displayed, as shown in Figure 5-6. In this page, specify the ID and name of the resource in the **Resource ID** and **Resource Name** edit boxes, respectively. Now, choose the **Next** button; the **Resource Type** page will be displayed, as shown in Figure 5-7.

Note
*If you choose the **Add** button for the first time in the **Resources** window, the **Information** message box will be displayed with the message that in order to add a resource, filter must be changed to display all resources. Choose the **OK** button; all resources will be displayed in the **Resources** window and the **New Resource Wizard** will be displayed, refer to Figure 5-6.*

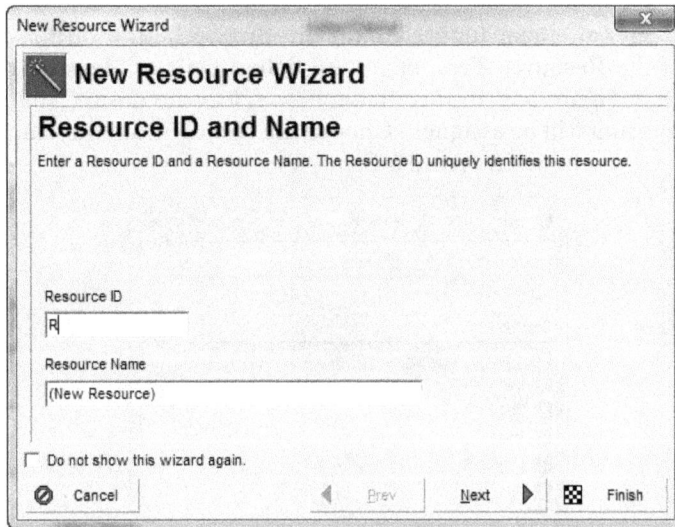

*Figure 5-6 The **Resource ID and Name** page of the **New Resource Wizard***

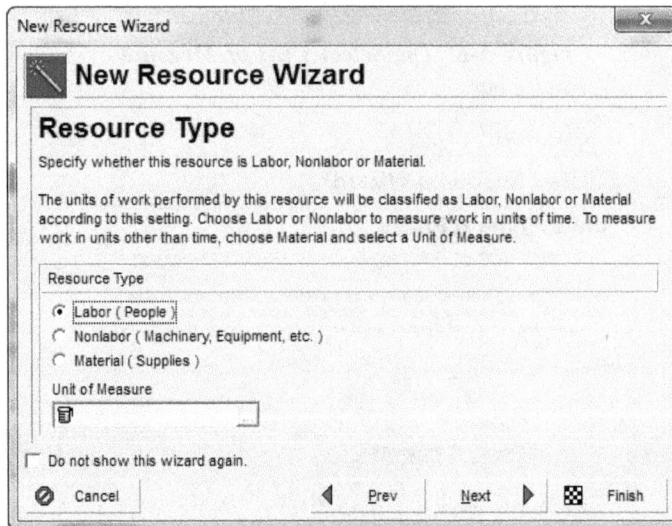

*Figure 5-7 The **Resource Type** page of the **New Resource Wizard***

In the **Resource Type** page, you are required to assign the resources type as Labor, Nonlabor, or Material by selecting the respective radio button. If the resource type is either Labor or Nonlabor, then it will be measured in units of time. In case, you select the **Material** radio button, then the **Unit of Measure** field below it will be enabled and you will be prompted to enter the unit of measurement. However, if you want to create your own type of unit of measurement, choose the **Admin Categories** option from the **Admin** menu; the **Admin Categories** dialog box will be displayed. In this dialog box, choose the **Unit of Measure** tab and then choose the **Add** button; the **Unit** under the **Unit Abbreviation** column will be displayed. Now, click to assign the desired unit name and then choose the **Close** button to close the dialog box.

Now, to specify the measurement unit, click on the Browse button on the right of the **Unit of Measure** field of the **Resource Type** page; the **Select Unit of Measure** dialog box will be displayed, as shown in Figure 5-8. In this dialog box, select the desired unit and then choose the **Select** button; the unit will be assigned. Choose the **Next** button; the **Units /Time & Prices** page will be displayed, as shown in Figure 5-9.

Figure 5-8 The Select Unit of Measure dialog box

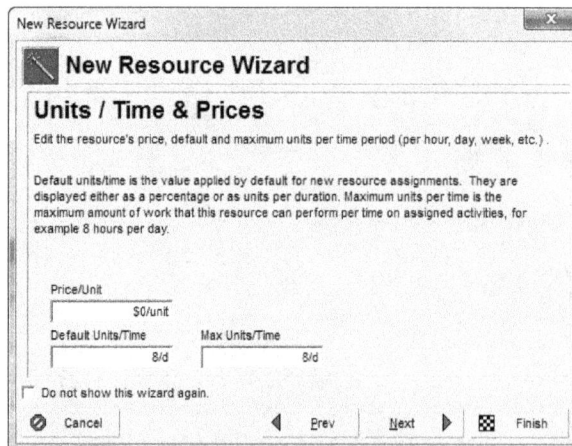

Figure 5-9 The Units/Time & Prices page of the New Resource Wizard

In this page, you need to specify the resource price, default units/time and maximum units/ time period. To enter these values, click in the respective edit boxes and set the required values. Choose the **Next** button; the **Phone & E-Mail** page will be displayed. In this page, you can save the personal details of the resources/employees. To enter the phone number and the email address of the resource, enter the details in the **Office Phone** and **E-Mail Address** edit boxes, respectively. Next, choose the **Next** button; the **Resource Calendar** page will be displayed.

Note

*In the **Resource Type** page, if you select the labor or non-labor resource type then after the **Units/Time & Prices** and the **Phone & E-Mail** page, the **Roles** page will be displayed, as shown in Figure 5-10. In the **Roles** page, you need to assign the role type according to the resources type selected that is either labor or non-labor. Choose the **Assign** button; the **Assign Roles** dialog box will be displayed. In this dialog box, change the display to all resources and assign the required role. Choose the **Assign** button and then close the dialog box. On doing so, a column gets added to the **Roles** page and you are required to enter necessary details under the **Role Name** and **Proficiency** columns. If you want the role to be primary, select the check box under the **Primary Role** column and then choose the **Next** button; the **Resource Calendar** page will be displayed.*

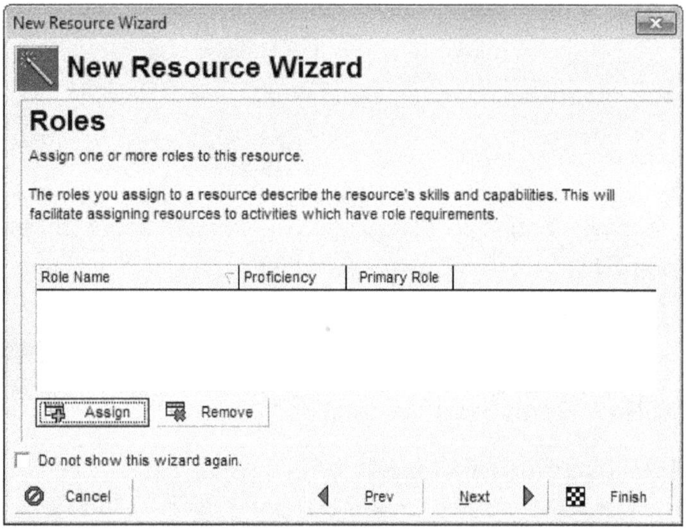

*Figure 5-10 The **Roles** page of the **New Resource Wizard***

As you have already learnt that you need to assign a calendar to resource for proper scheduling. You can assign a calendar from the existing pool of shared calendars or can create a new one. To assign an existing calendar, select the **Select an existing calendar** radio button in the **Resource Calendar** page and then choose the **Next** button; the **Select Existing Calendar** page will be displayed, as shown in Figure 5-11. This page displays the calendar which is assigned to the role. If you want to edit the existing calendar, choose the **View/Edit** button; the **Calendar** dialog box will be displayed. Make changes in the existing calendar and choose the **OK** button. If you want to replace the existing calendar, choose the Browse button in the **Calendar Name** field; the **Select Resource Calendar** dialog box will be displayed. In this dialog box, select the desired type of calendar and then choose the **Select** button. Then, choose the **Next** button; the **Auto Compute Actuals** page will be displayed.

To create a separate resource calendar, choose the **Create a new calendar for this resource** radio button in the **Resource Calendar** page and then choose the **Next** button; the **View and Edit Calendar** page will be displayed. In this page, choose the Browse button in the **Calendar Name** field; the **Select Resource Calendar** dialog box will be displayed. In this dialog box, select the desired type of calendar and then choose the **Select** button. Now, choose the **Next** button; the **Auto-Compute Actuals** page will be displayed, as shown in Figure 5-12.

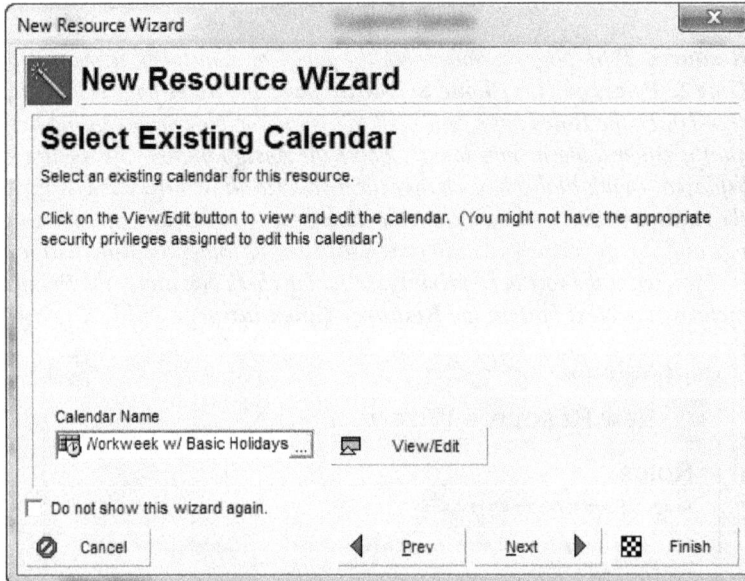

Figure 5-11 The Select Existing Calendar page of the New Resource Wizard

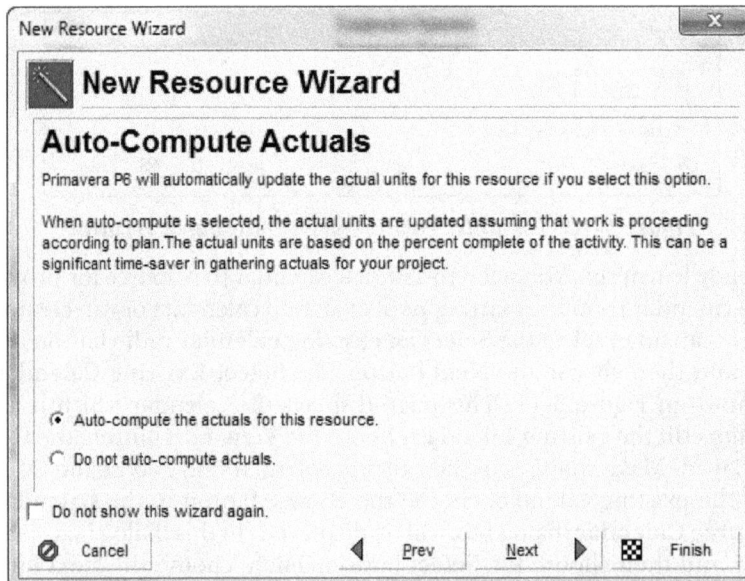

Figure 5-12 The Auto-Compute Actuals page of the New Resource Wizard

The radio buttons in the **Auto-Compute Actuals** page are used to specify whether the resources will auto compute or not. If you select the **Auto-Compute the actuals for this resource** radio button, the actual units for the resources will be updated automatically. On selecting the **Do not auto-compute actuals** radio button, the actual units will not be updated. After selecting any one of the radio buttons, choose the **Next** button; the **Progress Reporter Setup** page will be displayed. The progress reporter is an application that helps the user or the staff members to access the activities and report their status and time. If you want to run this application, then select the **Yes, I want to set up Progress Reporter** radio button. Otherwise, select the **No, I will not set up Progress Reporter at this time** radio button. Now, choose the **Next** button; the **Progress Reporter Login** page will be displayed. In this page, choose the Browse button in the **User Login** field; the **Select User Login Name** dialog box will be displayed. In this dialog box, select the user login name that will be used by the resource to log in to the Progress Reporter and then choose the **Select** button; the user login name will be assigned and the dialog box will be closed. After assigning the Progress Reporter Login name, choose the **Next** button; the **Congratulations** page will be displayed with the message that the resource is added, refer to Figure 5-13. Choose the **Finish** button; the resource is added with complete details.

Note
*The **Progress Reporter Login** page will not be displayed if the **No, I will not set up Progress Reporter at this time** radio button is selected in the **Progress Reporter Setup** page.*

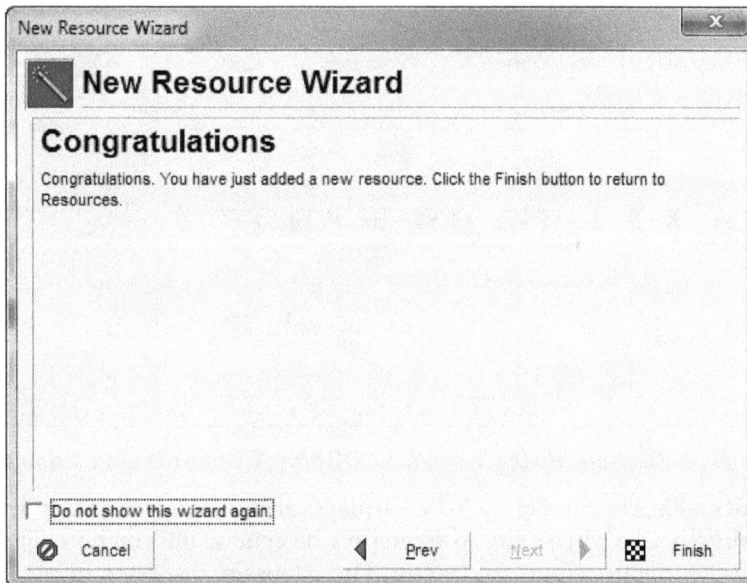

*Figure 5-13 The **Congratulations** page of the **New Resource Wizard***

ROLES
While planning a project, you may need to add the specific types of resources that are required for each type of activity. For example, you may have a number of project managers, but before a project is initiated, you may not know which project manager will be available for the new project. These resource types are represented by roles.

Generally, roles are defined for the enterprise, and not for a specific project. To add roles in an enterprise, choose **Enterprise > Roles** option from the menubar. On doing so, the **Roles** dialog box will be displayed with the **Display: Current Project's Roles** area, as shown in Figure 5-14. In this area, no role will be displayed if there are no roles assigned to the project. You can change this display by clicking on the **Display** drop-down; a menu will be displayed. Choose **Filter By > All Roles** option from the menu displayed. In the **Display: All Roles** area, you can add your own role type or you can assign the existing roles from the display. Use the Command bar to add, delete, or cut the role. Below the **Roles** window, the **Roles Details** table will be displayed. You can hide and unhide the details table by using the **Roles Details** option from the **Display** drop-down.

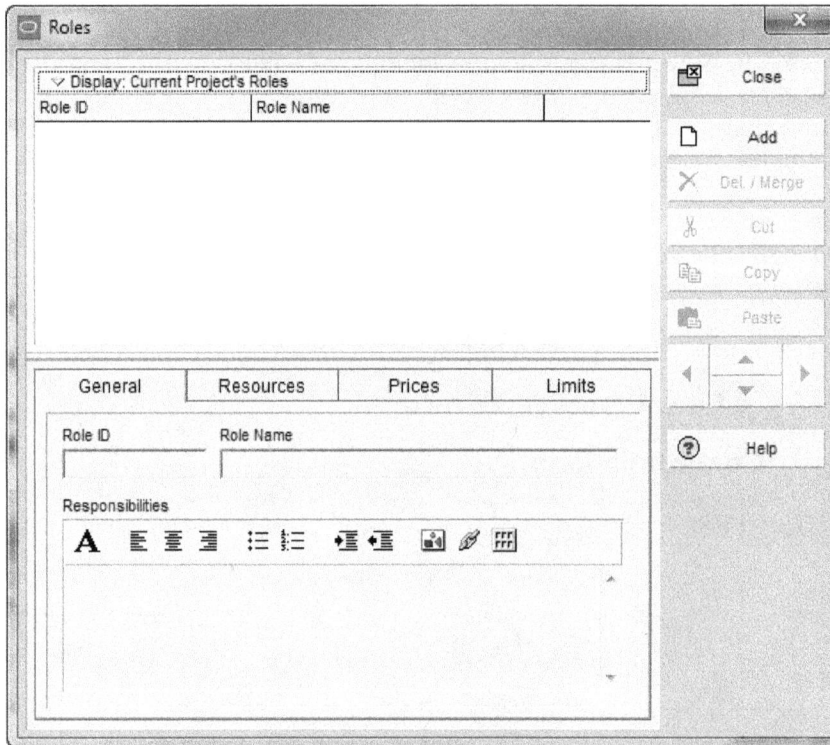

*Figure 5-14 The **Roles** dialog box with the **Display: Current Project's Roles** area*

The **Roles Details** table, refer to Figure 5-14, displays various tabs such as **General**, **Resources**, **Prices**, and **Limits** tab. Using these tabs, you can provide general information, select the resource to assign the role, set the price/unit, and so on. These tabs are discussed next.

General Tab
In this tab, you can specify the role name and role ID in the **Role Name** and **Role ID** edit boxes, respectively. A rich-text description of the role's responsibilities can be added in the **Responsibilities** area.

Resources Tab
The **Resources** tab shows all resources that are currently associated with the role. Similar to resources, roles can also be assigned a proficiency level from **1-Master** to **5-Inexperience**. You

can see whether each resource has this role as its primary role. In the **Resources** tab, to assign roles to the resources, choose the **Assign** button; the **Assign Resources** dialog box will be displayed. In this dialog box, select the resource from the **Display: All Resources** area to be assigned to the role.

Prices Tab

The **Prices** tab allows you to assign rate to a role. It is used for cost calculations. When you assign a resource to an activity, you can choose the rate type based on the role or the specific resource.

Limits Tab

The **Limits** tab, refer to Figure 5-15, enables you to set a limit on the number of units that a resource can work in a given unit of time. Generally, the unit for time is set to hours per day. The limit is set according to effective dates so that different amounts can be applied for different time periods.

The **Limits** tab serves similar purpose for role as the **Units and Prices** tab serves for the resources. Note that the time assigned in the **Max Units/Time** column of the **Limits** tab are the maximum hours per day for the maximum units of resource to work on projects. The units of resources may not necessarily be the same as the total hours of work per day.

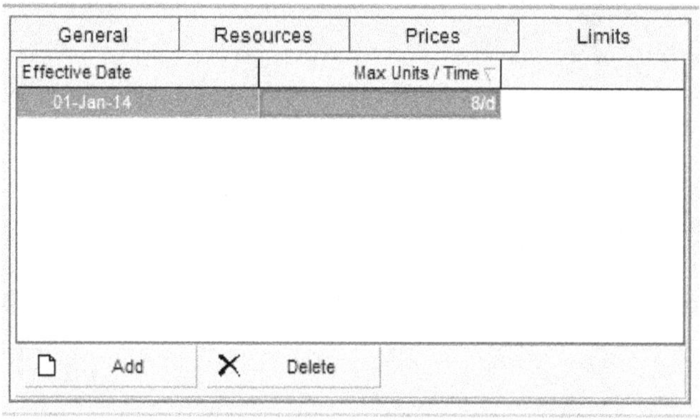

General	Resources	Prices	Limits
Effective Date		Max Units / Time	
01-Jan-14		8/d	

| | | | |
| Add | | X Delete | |

Figure 5-15 *The **Limits** tab of the **Roles Details** table*

ASSIGNING RESOURCES

After creating resources and roles, they must be assigned to an activity. After assigning the resources and roles, you must run a schedule. A schedule in which all relevant activities have roles or resources assigned is called a resource loaded schedule.

To assign resources and roles to an activity, you need to be in the **Activities** window. To invoke this window, choose the **Activities** option from the **Project** menu; the **Activities** window will be displayed. Alternatively, choose the **Activities** tab from the Directory bar. In the **Activities** window, ensure that the **Resources** tab is chosen from the **Activity Details** table. In the **Resources** tab, four buttons **Add Resource**, **Add Role**, **Assign by Role**, and **Remove** will be displayed at the bottom. These buttons are used to assign resources and roles to an activity. The functions of these buttons are explained next.

The **Add Resource** button allows you to assign resource to an activity. On [🖳 Add Resource] choosing the **Add Resource** button; the **Assign Resources** dialog box will be displayed. In this dialog box, ensure that the **All Resources** option is selected in the **Display** drop-down. As a result, all the resources present in the module will be displayed. Select the desired resource from the displayed resources list. When you add a resource, the role chosen is the resource's primary role by default, but this can be altered as needed. To make the resource as the primary resource, select the check box in the **Primary Resource** column of the corresponding resource. If the **Primary Resource** column is not displayed, you can customize the **Resources** column. To do so, right-click anywhere in the **Resources** tab and choose the **Customize Resource Columns** option; the **Resource Assignments Columns** dialog box will be displayed. In this dialog box, choose the options from the **Available Options** area and shift them to the **Selected Options** area using the **Add to list** button.

The **Add Role** button is used to assign roles to the resources that are required to [🖳 Add Role] complete an activity. Choose the **Add Role** button; the **Assign Roles** dialog box will be displayed. In this dialog box, ensure that the **Display: All Roles** area is displayed. If not then choose **Filter By > All Roles** option from the **Display** drop-down. In this list, select the role type which is required to be assigned to the resources. When you assign a role to an activity, the default values of the prices/unit and units/time are displayed under the **Price/Unit** and **Units/Time** columns, respectively.

The **Assign by Role** button allows to assign the roles to an activity when no [🖳 Assign by Role] resource is added yet. To assign roles to an activity, choose the **Assign by Role** button; the **Assign Resources By Role** dialog box will be displayed. By default, the **Current Project's Resources** option is selected in the **Display** drop-down. You can choose the **Filter By** option from the **Display** drop-down to invoke the **Filter By** dialog box. In this dialog box, you can select the **All Resources** radio button or the **Current Project's Resource** radio button to display the resources accordingly. The **Select Filter** area displays the following options: **All Roles Required**, **Staffed Roles**, **Unstaffed Roles Required**, **Unstaffed Roles with Required Proficiency** and so on.

Note

*If no role is not assigned to an activity then on choosing the **Assign by Role** button, the **Primavera P6** message box will be displayed. This message box informs you that the selected activity does not have any role assigned. Choose the **OK** button and then assign the role to the activity.*

In the **Select Filter** area, the **All Roles Required** option is selected by default and it shows every available resource within a project has the roles specified in the activity. Note that many resources may appear multiple times if they have more than one role. The **Staffed Roles** option is used to display only those roles to which a specific resource has been assigned. The **Unstaffed Roles Required** option is used to displays all resources that are capable of performing the needed roles but have not yet been assigned with any work. The **Unstaffed Roles with Required Proficiency** option allows you to assign a role according to the proficiency level. While assigning a role to an activity, you can choose the skill level required for that role.

RESOURCE CURVES

When resources are assigned to an activity, you need to choose the resource units or costs that are allocated according to time. A resource curve enables you to determine how the effort and cost of a resource is spread over time. This helps in calculating the expected cost for labor or the expected consumption of a material resource.

You can represent the resource curve in the **Curve** column of the **Resources** tab of the **Activity Details** table. If the **Curve** column is not displayed in the **Resource tab**, you can customize the columns of the **Resources** tab using the **Resource Assignment Columns** dialog box. To assign curve to the resource, you need to click on the **Curve** column; the **Select Curve** dialog box will be displayed, as shown in Figure 5-16. In this dialog box, select the required resources curve and then choose the **Select** button; the curve will be assigned to the resource and the dialog box will be closed.

However, to add your own type of curve, choose the **Resource Curves** option from the **Enterprise** menu; the **Resource Curves** dialog box will be displayed, as shown in Figure 5-17. This dialog box allows you to add, modify and delete curves. These processes are discussed next.

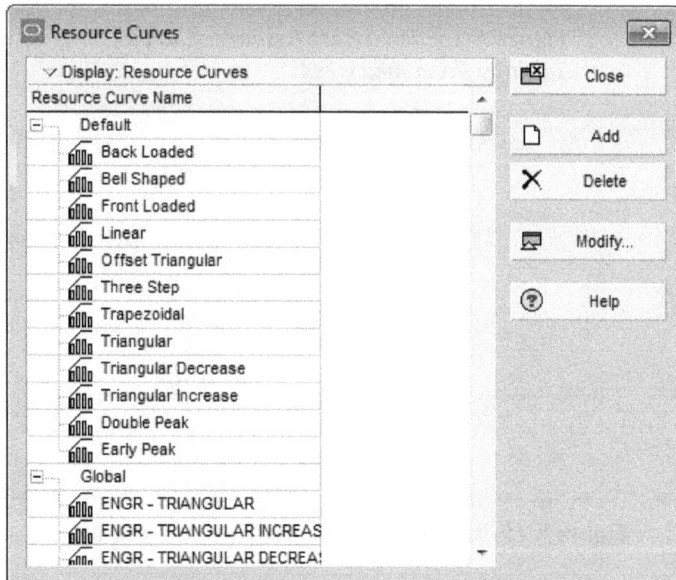

*Figure 5-16 The **Select Curve** dialog box*

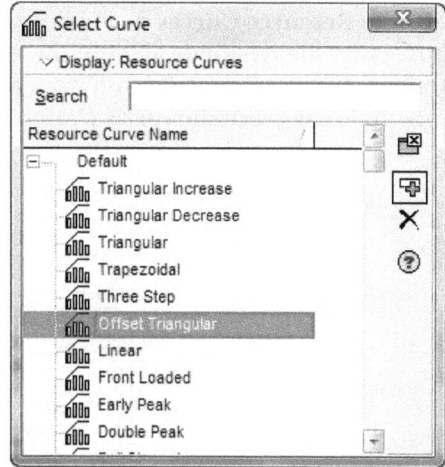

*Figure 5-17 The **Resource Curves** dialog box*

Choose the **Add** button to add a new type of resource curve; the **Select Resource Curve To Copy From** dialog box will be displayed, as shown in Figure 5-18. In this dialog box, select the required resource curve that can be taken as a reference resource type and then choose the **Select** button; the dialog box will be closed and the resources curve with the **New Resource Curve** name will be added in the **Resource Curve Name** area. You can edit its name by clicking on it and name it as required. Now, to modify the copied resource curve type, choose the **Modify** button; the **Modify Resource Curves** dialog box will be displayed, as shown in Figure 5-19. In this dialog box, enter the required values in the edit boxes displayed at the bottom of the dialog box and then choose the **Prorate** button from the Command bar to update the curve. After making the required changes in the curve, choose the **OK** button; the dialog box will be closed.

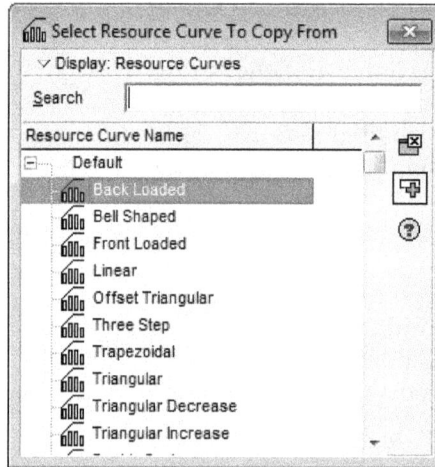

Figure 5-18 The Select Resource Curve To Copy From dialog box

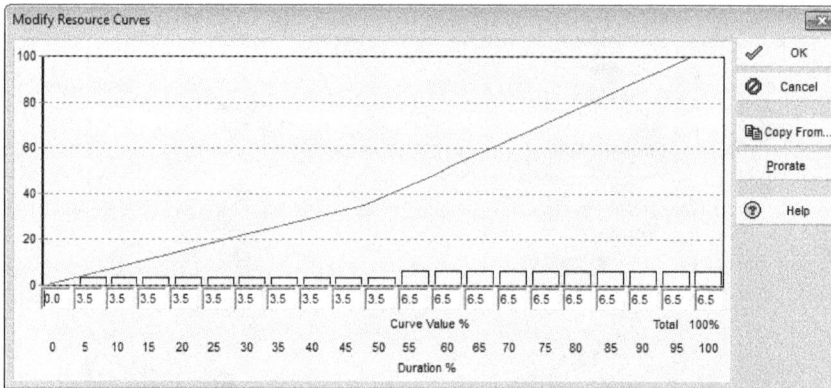

Figure 5-19 The Modify Resource Curves dialog box

TUTORIALS

To perform the tutorials of this chapter, you need to complete Tutorial 1 and Tutorial 2 of Chapter 2.

To perform this tutorial, you need *c04_Primavera_P6_tut.zip* file. You can download this file from *www.cadcim.com* using these steps.

1. To download the file, browse to *Textbooks > Civil/GIS > Primavera P6 > Exploring Oracle Primavera P6 v7.0*. Next, select the *c04_Primavera_P6_tut.zip* file from the **Tutorial Files** drop-down list. Choose the corresponding **Download** button to download the data file.

2. Now, save and extract the downloaded folder to the following location:
    C:/PM6

3. Import the *c04_CONS_Home_tut02* file from the extracted folder to the Construction EPS created in Tutorial 1 of Chapter 2.

Tutorial 1   Resources & Roles

In this tutorial, you will create resources and then assign resources to them for the **Home Construction** project.   **(Expected time: 30 min)**

The following steps are required to complete this tutorial:

a. Open the **Resources** window.
b. Customize the **Resources** column.
c. Add resources to the project.
d. Assign prices to resources.
e. Export the project.

Opening the Resources Window

1. Open Primavera P6 and invoke the **Projects** window by choosing the **Projects** option from the **Enterprise** menu in the menubar.

2. In this window, expand the **CADCIM** EPS and then select the **Home Construction** project placed under the **Construction** EPS node and then right-click; a shortcut menu is displayed.

3. Choose the **Open Project** option from this menu; the **Activities** window with the **Home Construction** project is displayed.

4. Choose the **Resources** option from the **Enterprise** menu; the **Resources** window is displayed.

Customizing the Resources Columns

1. In this window, choose the **All Resources** option from the **Filter By** cascading menu in the **Display** drop-down, as shown in Figure 5-20, to display all resources.

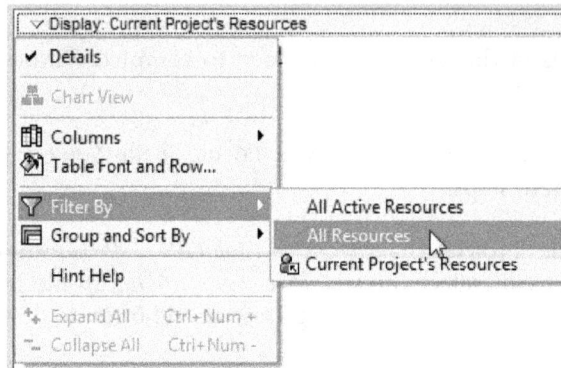

Figure 5-20 *Choosing the **All Resources** option from the **Filter By** cascading menu*

2. Right-click anywhere in the **Resources** window; a menu is displayed.

3. Choose the **Columns > Customize** option; the **Columns** dialog box is displayed.

4. Using the CTRL key, select the **Unit of Measure** and **Default Units / Time** options from the **Selected Options** area and shift them to the **Available Options** area using the **Remove from list** button.

5. Ensure that the **General** node is expanded in the **Available Options** area.

6. Using the CTRL key, select the **Max Units/Time** and **Price/Unit** options from the **Available Options** area and shift them to the **Selected Options** area using the **Add to list** button.

7. In the **Selected Options** area, arrange the displayed options, as shown in Figure 5-21.

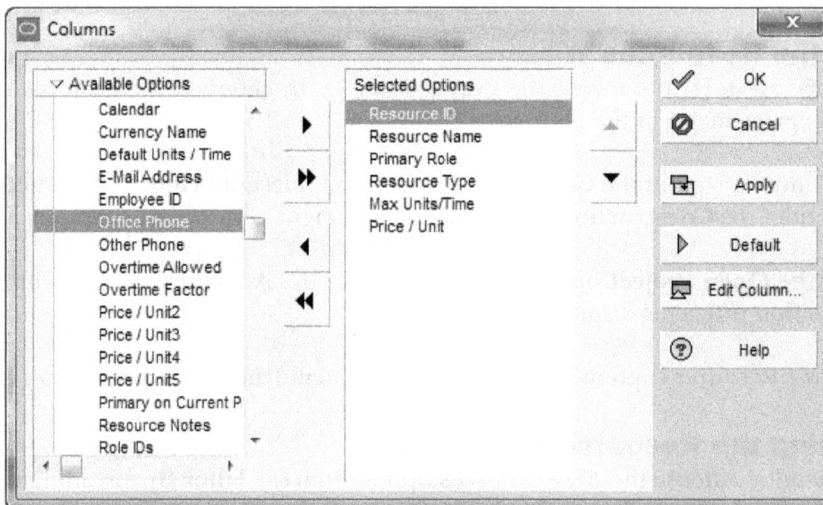

Figure 5-21 *The options arranged in the **Selected Options** area*

8. Choose the **OK** button; the dialog box is closed.

Adding the Resources

1. Select the first resource from the **Resources** window and choose the **Add** button from the Command bar; the **New Resource Wizard** is displayed.

2. In this wizard, enter **STV** in the **Resource ID** edit box and **Steve Jones** in the **Resource Name** edit box.

3. Choose the **Next** button; the **Resource Type** page is displayed.

4. In this page, ensure that the **Labor (People)** radio button is selected in the **Resource Type** area.

5. Choose the **Next** button; the **Units / Time & Prices** page is displayed.

6. In this page, enter **$90.00/h** in the **Price/Unit** edit box and **1/d** in the **Max Units / Time** edit box.

7. Choose the **Next** button twice; the **Roles** page is displayed.

8. In the **Roles** page, choose the **Assign** button; the **Assign Roles** dialog box is displayed.

9. Ensure that the **All Roles** option is selected in the **Display** drop-down. If not, then choose **All Roles** option from the **Filter By** cascading menu of the **Display** drop-down.

10. Select the **Engineer** option from the **Role Name** column, as shown in Figure 5-22, and then choose the **Assign** button. Close the dialog box; the **Roles** page is displayed.

11. Click under the **Proficiency** column; a drop-down list is displayed. Select the **1-Master** option from the drop-down list.

12. Choose the **Next** button; the **Resource Calendar** page is displayed.

13. Choose the **Next** button twice; the **Auto-Compute Actuals** page is displayed. Choose the **Next** button; the **Progress Reporter Setup** page is displayed.

14. In this page, select the **No, I will not set up Progress Reporter at this time** radio button.

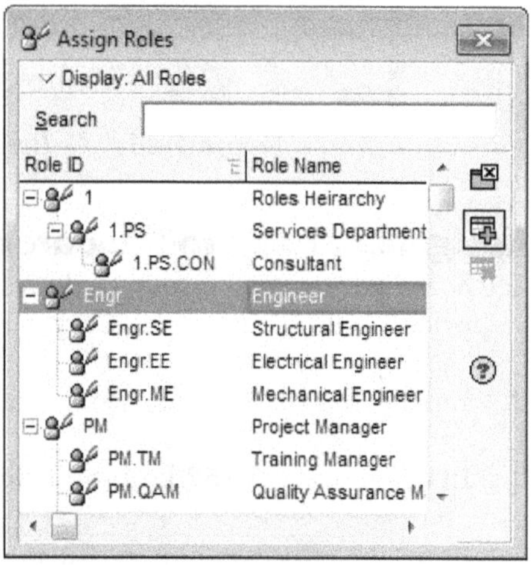

Figure 5-22 The Engineer option selected in the Assign Roles dialog box

15. Choose the **Next** button; the **Congratulations** page is displayed and then choose the **Finish** button; the resource is created with the role.

16. Choose the **Left** button from the Command bar and then choose the **Up** button to set the created resource at the top.

17. Repeat the procedure followed in steps 1 through 15 and enter the details, as shown in Table 5-1, in the pages of the wizard to create other resources.

Note that, you can select the **3-Skilled** option for other resources from the **Proficiency** column of the **Roles** page. Also, select the **Material** radio button for the **Concrete** resource type and assign the **Cubic Yards** option as a **Unit of Measure** in the **Resource Type** page.

Table 5-1 *The necessary details that need to be filled while adding resources*

Resource ID	Resource Name	Resource Type	Price/Unit	Max Units/Time	Role Name
DMR	Daniel Muller	Labor	$55.00/h	8/d	Foreman
RCS	Richard Collins	Labor	$40.00/h	32/d	HVAC
Surveyor	Land Surveyor	Labor	$65.00/h	16/d	Sitework
Siding	Siding Contractor	Labor	$47.00/h	16/d	Sitework
CON	Concrete	Material	$10/cu yd	8/d	No role

After creating the resources, arrange the resources, as shown in Figure 5-23.

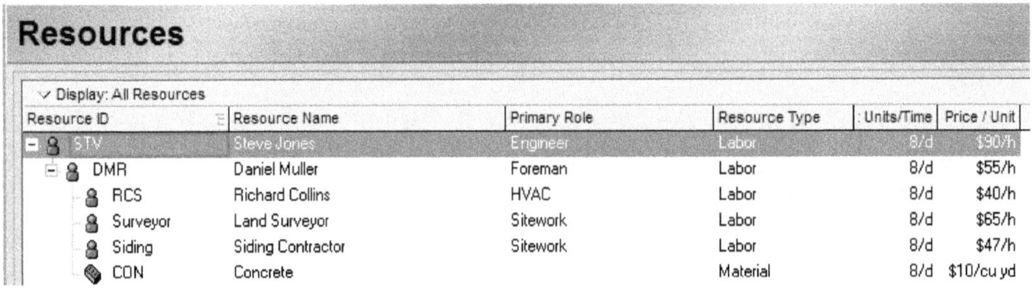

Figure 5-23 *The **Resources** window with the added resources*

Assigning Prices to Resources

1. Some pre-existing resources do not have prices. Therefore, you need to enter the resource price. To do so, select the **Carptr** resource ID from the **Resource ID** column.

2. Ensure that the **Units & Prices** tab is chosen in the **Resources Details** table.

3. In this tab, enter **$52/h** in the **Price/Unit** column.

4. Enter the effective date for Carptr as **12-Dec-15** under the **Effective Date** column.

5. Similarly, assign prices and the effective date to other resources, as shown in Table 5-2.

Table 5-2 *The details required to be filled while adding resources*

Resource ID	Resource Name	Price/Unit	Max Units/Time	Effective Date
Carptr	Carpentry Crew	$52.00/h	32/d	12-Dec-15
ENGR Labor	Engineer Labor	$18.00/h	32/d	13-Oct-15
Steel	Steel Crew	$49.00/h	48/d	13-Oct-15
Plum	Plumbing Crew	$58.00/h	8/d	16-Oct-15
Plaster	Plaster Crew	$49.00/h	8/d	5-Nov-15
Mason	Mason	$10.00/h	8/d	13-Oct-15
RCS	Richard Collins	$60.00/h	8/d	6-Jan-16
Electr	Electrical Crew	$52.00/h	1/d	1-Dec-15
Glaz	Glass Crew	$70.00/h	8/d	3-Nov-15
Landsc	Landscape Crew	$72.00/h	8/d	7-Nov-15

Exporting Project

You need to export projects to save the project file for future reference. Ensure that the project which is to be exported is opened in Primavera P6.

1. To export a project, choose the **Export** option from the **File** menu; the **Export** wizard with the **Export Format** page is displayed.

2. In this page, ensure that the **Primavera PM/MM - (XER)** radio button is selected and then choose the **Next** button; the **Export Type** page is displayed.

3. In this page, ensure that the **Project** radio button is selected and then choose the **Next** button; the **Projects To Export** page with the **Home Construction** project is displayed.

4. Choose the **Next** button; the **File Name** page is displayed.

5. In this page, choose the Browse button from the **File Name** edit box; the **Save File** dialog box is displayed.

6. In this dialog box, browse to the *C:\PM6* and then create a sub-folder with the name **c05**.

7. Open the created folder and save the file with the name *c05_CONS_HOME_tut01* and then choose the **Save** button.

8. Choose the **Finish** button; the **Primavera P6** message box is displayed with the message that the export was successful.

9. Choose the **OK** button; the file is exported.

Tutorial 2   Resource Assignments

In this tutorial, you will assign resources to activities in the **Home Construction** project.
(Expected time: 30 min)

The following steps are required to complete this tutorial:

a. Open the **Activities** window.
b. Customize the Resources Column.
c. Assign Resources to activity.

Opening the Activities Window

1. Open Primavera P6 and invoke the **Projects** window by choosing the **Projects** option from the **Enterprise** menu in the menubar.

2. In this window, select the **Home Construction** project placed under the **Construction** EPS node and right-click; a shortcut menu is displayed.

3. Choose the **Open Project** option from this menu; the **Activities** window with the activities of the **Home Construction** project is displayed.

Customizing the Resource Columns

1. In the **Activities** window, ensure that the **Activity Details** table is displayed with the **Resources** tab chosen.

2. Right-click in the white space area of the **Resources** tab; the **Customize Resource Columns** option is displayed.

3. On selecting this option, the **Resources Assignment Columns** dialog box is displayed.

4. In this dialog box, select the **Curve, Cost Account, Actual This Period Units,** and **Actual Units** options from the **Selected Options** area and choose the **Remove from list** button.

5. In the **Available Options** area, expand the **General** node and select the **Budgeted Units/ Time** option. Next, choose the **Add to list** button to move the option to the **Selected Options** area.

6. Similarly, expand the **Duration** node and select the **Original Duration** option. Next, choose the **Add to list** button to move the option to the **Selected Options** area.

7. Choose the **OK** button; the **Columns** dialog box is closed and the columns are customized.

Assigning Resources

1. In the **Activities** window, select the **Project Management** activity and then choose the **Add Resource** button from the **Activity Details** table; the **Assign Resources** dialog box is displayed, refer to Figure 5-24.

2. Select the **Daniel Muller** option from the list in the **Assign Resources** dialog box and choose the **Assign** button. Next, close the dialog box.

3. In the **Activity Details** table, ensure that the **Budgeted Units/Time** of this resource is 8/d.

4. Select the **Survey and mark out site** activity from the **Activities** Window.

5. Choose the **Add Resources** button; the **Assign Resources** dialog box is displayed.

6. Select the **Surveyor** option from the **Assign Resources** dialog box, as shown in Figure 5-24.

7. Choose the **Assign** button and then close the dialog box using the **Close** button.

8. Enter **8/d** in the **Budgeted Units/Time** column of the **Activity Details** table.

9. Similarly, select the **Site Demarcation** activity and then select the **Add Resource** button; the **Assign Resources** dialog box will be displayed.

Figure 5-24 The Surveyor option selected in the Assign Resources dialog box

10. Select the **Surveyor** option from this dialog box and then choose the **Assign** button.

11. Now, select the **Excavation** activity from the **Activities** window and choose the **Add Resource** button.

12. In this dialog box, select the **Crane** option and then choose the **Assign** button.

13. Enter **8/d** in the **Budgeted Units/Time** column and ensure that the price is already entered in the **Price/Unit** column.

14. Now, select the **Grade Site** activity and choose the **Add Resource** button; the **Add Resources** dialog box will be displayed.

15. Enter **Oper** in the **Search** edit box; three options are displayed in the **Resource ID** and **Resource Name** columns, as shown in Figure 5-25.

16. Select **Op Eng** option in the **Resource ID** column and choose the **Assign** button. Then, close the dialog box.

17. Enter **8/d** in the **Budgeted Units/Time** column of the **Resources Details** table.

18. Repeat the procedure followed in previous steps and assign the resources. Table 5-3 shows the necessary details required to assign resources.

   Note that after assigning resources, if the original duration is not displayed in the **Original Duration** column of the **Activity Details** table, you need to enter the value as mentioned in the **Activities** window for the respective option. On doing so, the Budgeted Units will be calculated automatically.

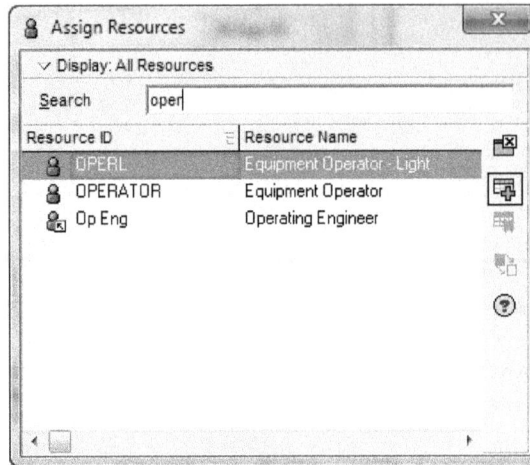

Figure 5-25 *Options displayed on entering* **Oper** *in the* **Search** *edit box*

Table 5-3 *The details required to assign the resources to the activity*

Activity ID	Activity Name	Resources ID	Budgeted Units/Time
A1020	Project Management	DMS	8/d
A1030	Survey and Mark Out Site	Surveyor	8/d
A1040	Grade Site	OpEng	8/d
A1050	Laying of Foundation	ENGR Labor	8/d
		Steel	8/d
		Carptr	8/d
		CON	8/d
		Mason	8/d
A1060	Slab Plumbing	Plum	8/d
A1070	Pour and float slab concrete	CON	8/d
A1080	Exterior Walls	Mason, Plas	8/d
A1090	Interior Walls	Mason, Plas	8/d
A1100	Exterior Claddings	Mason, Plas	8/d
A1110	Truss Placings	Steel	8/d
A1120	Roof Sheetings	LABOR-S	8/d
A1130	Place paper and shingles	LABOR-S	8/d
A1140	HVAC Ducting	RCS	8/d
A1150	Install HVAC unit	RCS	8/d
A1160	Breaker Box and rough wire	Electr	8/d
A1170	Finish Wiring	Electr	8/d
A1180	Install rough plumbing lines	Plum	8/d
A1190	Placing Doors	Carptr	8/d
A1200	Doors Casings and Baseboards	Carptr	8/d
A1210	Kitchen Cabinets	Carptr	8/d
A1220	Place Windows	Glaz, Carptr	8/d
A1230	Window Sidings	Carptr	8/d
A1240	Tree Plantings	Landsc	8/d

Exporting Project

You need to export projects to save the project file for future reference. Ensure that the project which is to be exported is opened in Primavera P6.

1. To export a project, choose the **Export** option from the **File** menu; the **Export** wizard with the **Export Format** page is displayed.

2. In this page, ensure that the **Primavera PM/MM - (XER)** radio button is selected and then choose the **Next** button; the **Export Type** page is displayed.

3. In this page, ensure that the **Project** radio button is selected and then choose the **Next** button; the **Projects To Export** page with the **Home Construction** project is displayed.

4. Choose the **Next** button; the **File Name** page is displayed.

5. In this page, choose the Browse button from the **File Name** edit box; the **Save File** dialog box is displayed.

6. In this dialog box, browse to the *C:\PM6\c05* and save the file with the name **c05_CONS_HOME_tut02** and then choose the **Save** button.

7. Choose the **Finish** button; the **Primavera P6** message box is displayed with the message that the export was successful.

8. Choose the **OK** button; the file is exported.

Self-Evaluation Test

Answer the following questions and then compare them to those given at the end of this chapter:

1. You can enter the general information of resources in the _____ tab.

2. You can enter the date for resource's maximum units/time in the _____ tab.

3. You can assign the proficiency of resources in the _____ page.

4. Resources are defined as the stock of labor, non labor, and material. (T/F)

5. Resources are one time expenditure for non reusable items. (T/F)

6. You can add resources from the **Resources Details** table. (T/F)

7. You can enter the number of resources in the **New Resource** wizard. (T /F)

8. You can enter the complete details of the resources in the **Resources Details** table. (T/F)

Review Questions

Answer the following questions:

1. Which of the following tabs enables you to enter the overtime factor?

    (a) **Roles**   (b) **Details**
    (c) **Codes**   (d) **Notes**

2. Which of the following tabs allows you to enter the role ID and the name of the role?

    (a) **Notes**   (b) **Units & Prices**
    (c) **General**   (d) **Progress Reporter**

3. Which of the following menus enables you to display the **Resource Assignments** window?

    (a) **Edit**   (b) **Admin**
    (c) **Enterprise**   (d) **Project**

4. Resources and roles can be assigned only when they are created. (T/F)

5. The **Roles Details** table is displayed by default in the **Roles** dialog box. (T/F)

6. The Resource curve displays the resource units and costs. (T/F)

7. The **Information** message box is displayed whenever the **Add** button is chosen. (T/F)

8. The **Unit of measure** edit box is enabled on selecting the **Nonlabor** radio button. (T/F)

EXERCISES

Exercise 1

In this exercise, you will create roles and resources and then assign the created roles to resources. Next, you will assign the resources to the **Hospital Building** project.

(Expected time: 30 min)

Hints:
1. Open the Hospital Building Project.
2. Invoke the **Roles** dialog box to add the roles, as shown in Table 5-4.
3. Open the **Resources** window and add resources with the Jackey resources as the main head, as shown in Table 5-4.
3. Export the file with the name *c05_CN_BNQT_ex01*.

Table 5-4 The necessary details that need to be filled while adding resources

Resource ID	Resource Name	Resource Type	Price/Unit	Max Units/Time	Role ID	Role Name
MS	Jackey	Labor	$17.00/h	8/d	MS	Mason
POP	POP Crew	Labor	$17.00/h	8/d	POP	POP
Painter-1	Painter Head	Labor	$20.00/h	8/d	Paint	Painter
Road	Road Labor	Labor	$17.00/h	8/d	Road	Road Labor
Labor-S-1	Skilled Labor	Labor	$17.00/h	8/d	Lab	Labor

Exercise 2

In this exercise, assign the newly created as well as the existing resources to the **Hospital Building** project.   **(Expected time: 30 min)**

Table 5-5 *The details which are required to assign resources to the project*

Activity ID	Activity Name	Resource Names
Project Milestone		
A1000	Project Start	
A1010	Project Complete	
Mobilisation & Initial Setting out		
A1020	Initial Mobilisation by Civil Contractor	Project Manager
Construction Activities		
Structural Shell		
A1030	Lower Basement Roof Slab	Concrete Crew, Steel Crew
A1040	Upper Basement Roof Slab	Steel Crew, Concrete Crew
A1050	Ground Floor Roof Slab	Concrete Crew, Steel Crew
A1060	First Floor Roof Slab	Steel Crew, Concrete Crew
A1070	Second Floor Roof Slab	Concrete Crew, Steel Crew
A1080	Third Floor Roof Slab	Steel Crew, Concrete Crew
A1090	LMR/Mumty	Concrete Crew, Steel Crew
A1100	Foundation	Truck Crane, Concrete Crew, Steel Crew
Civil Finishes & Interiors		
Lower Basement		
A1110	Brick Work Internal	Mason Head, Skilled Labor
A1120	Services in walls	Mason Head
A1130	Internal Plaster	Plaster Crew
A1140	Services in Ceiling	Skilled Labor
A1150	Door Window Subframes	Carpentry Crew
A1160	Tiling	Mason Head, Skilled Labor
A1170	POP Work	POP Crew
A1180	Flooring	Mason Head, Skilled Labor, Plaster Crew
A1190	False Ceiling	POP Crew
A1200	Paint (Base + First Coat)	Painter Head
A1210	Placing of Doors and Windows	Carpentry Crew
A1220	Fixed Furniture	Carpentry Crew
A1230	Final Paintings	Painter Head

The resource assignments for **Upper Basement**, **First Floor**, **Second Floor**, and **Third Floor** are same as that of **Lower Basement**.

Table 5-6 shows the resource assignments of LMR/Mumty, Other Services, Lifts & Elevators, External Developments, and so on.

Table 5-6 *The resources assigned to the project*

LMR/Mumty		
A1890	Brick Work Internal/External	Mason Head, Skilled Labor
A1900	Services in walls	Mason Head
A1910	Plaster Internal/External	Plaster Crew
A1920	Services in Ceiling	Skilled Labor
A1930	Door Window Subframes	Carpentry Crew
A1950	POP Work	POP Crew
A1960	Flooring	Mason Head, Skilled Labor, Plaster Crew
A1980	Paint (Base + First Coat)/External Paint	Painter Head
A1990	Placing of Doors and Windows	Carpentry Crew
A2010	Final Paintings	Painter Head
Other Services		
A2020	Electrical	Electrical Crew
A2030	Plumbing	Plumbing Crew
A2040	HVAC	HVAC Eng
Lifts & Elevators		
A2050	Delivery of Lifts	
A2060	Installation and Commisioning of Lifts	
External Developments		
A2070	Landscape and Horticulture	Landscaping Subcontractor
A2080	Roads and external gates	Road Labor
A2090	External Lighting and Signages	Electrical Crew
Handing Over		
A2100	Loose Furniture	Labor
A2110	Ready for user trial	

Answers to Self-Evaluation Test

1. **General**, 2. **Units & Prices**, 3. **Roles**, 4. T, 5. F, 6. F, 7. F, 8. T

Chapter 6

Risks and Issues, and Setting Baselines

Learning Objectives

After completing this chapter, you will be able to:

* *Add risks in a project*
* *Add thresholds in a project*
* *Generate issues in a project*
* *Set the baseline in a project*
* *Assign and maintain baseline*

INTRODUCTION

In the previous chapter, you learned about the features and details of the resource layout. You also learned to add resources and roles to a project and assign those resources to the activities in the **Activities** window.

In this chapter, you will learn about risks, adding risks, managing risks and so on in a project. Also, you will learn about the issues and thresholds in a project. In addition, you will learn to create baseline, assign baseline, modify a baseline and update the baseline with new data as well as manage the baseline according to the schedule of the project. In the forthcoming section, project risks are explained in detail.

RISKS

Risk is defined as a situation involving exposure to danger. Risks are used to assign values to events that may or may not happen in a project. Risks are conditional and imaginary situations, which if happen impact the project objectives. Objectives may include aspects like scope, schedule, cost, and quality. You can add and manage the risks in Primavera P6. To add a risk to a project, choose the **Risks** option from the **Project** menu; the **Project Risks** window will be displayed, as shown in Figure 6-1. Alternatively, you can choose the **Risks** button from the Directory bar to display the **Project Risks** window. In this window, choose the **Add** button from the Command bar; the risk will be added to the **Display** area. Ensure that the **Risks Details** table is displayed below the **Project Risks** window. Select the **Risk Details** option from the **Display** drop-down to display the **Risks Details** table.

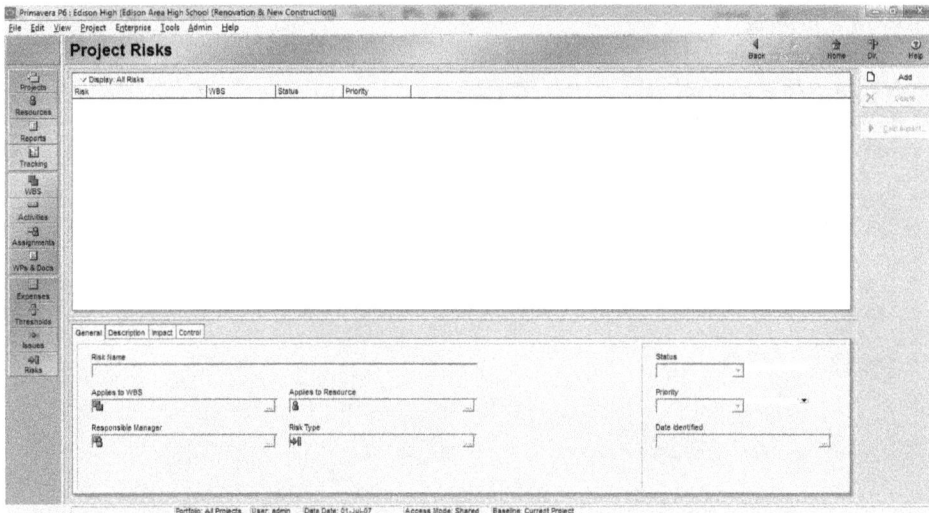

*Figure 6-1 The **Project Risks** window*

Risk Details Table

In the **Risk Details** table, four tabs are displayed: **General**, **Description**, **Impact**, and **Control**. Choose the **General** tab to enter the name of the risk in the **Risk Name** edit box. Next, in the **Applies to WBS** and **Applies to Resource** edit boxes specify the WBS element and the resource that will get affected, respectively. If you do not specify a resource, the module considers all the resources in the selected WBS. In the **Responsible Manager** edit box, you can specify the manager responsible for controlling the risk. Responsible managers are defined according to the Organizational Breakdown Structure (OBS). In the **Risk Type** edit box, specify risk type. Select the priority level from the **Priority** drop-down list. Priority level ranges from 1 to 5, where 1 indicates highest priority and 5 indicates lowest priority. To assign the risk, choose the Browse button in the **Risk Type** edit box and then select the type of risk. Next, choose the **Select** button; the risk type will be assigned. Assign status to the risk by selecting an option from the **Status** drop-down list.

In the **Description** tab, you need to enter the description of the selected risk. To enter the description, you can use HTML editing features such as text formatting, inserting pictures, copying and pasting information from other document files, and adding hyperlinks.

In the **Impact** tab, you can specify the impact date for the selected risk in the **Impact Date** edit box. You can also calculate exposure values according to labor units, non labor units, and expenses. Note that Primavera P6 module considers only those activities for the selected WBS/resource that are scheduled to start on or after the impact date. Enter the estimated number of labor and non labor, material, or estimated total cost of expenses in their corresponding edit boxes under the **Impact** column. Enter the percentage of probability in the **Probability** edit box. The exposure values will be calculated as impact times probability, refer to Figure 6-2.

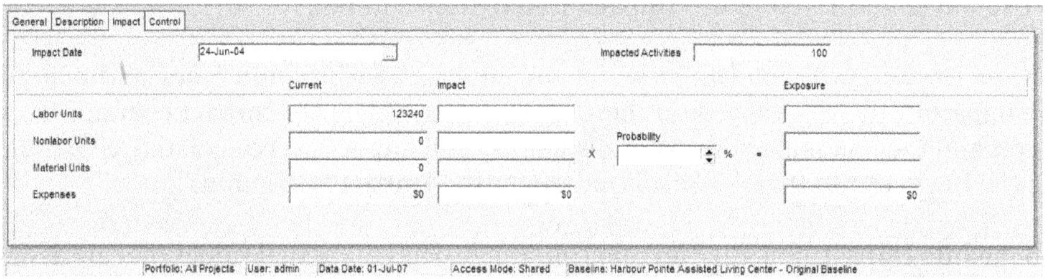

*Figure 6-2 The **Risk Details** table with the **Impact** tab chosen*

In the **Control** tab, you can type the description about the risk control plan. To add the description about the risk control plan, you can use HTML editing features that include text formatting, inserting pictures, copying and pasting information from other document files, and adding hyperlinks.

Calculating the Risk Impact

The risk impact is calculated by applying the exposure values of risk and the impact on the schedule, cost, and duration of a project using the top-down estimation method. The impact values calculated are applied to the activities that finish on or after the risk impact date. Note that these values are not valid for completed, locked, or milestone activities.

To calculate the risk impact, ensure that the impact values are assigned using the **Impact** tab. Now, choose the **Calc Impact** button from the Command bar; the **Risk Impact: (New Risk)** dialog box will be displayed, as shown in Figure 6-3.

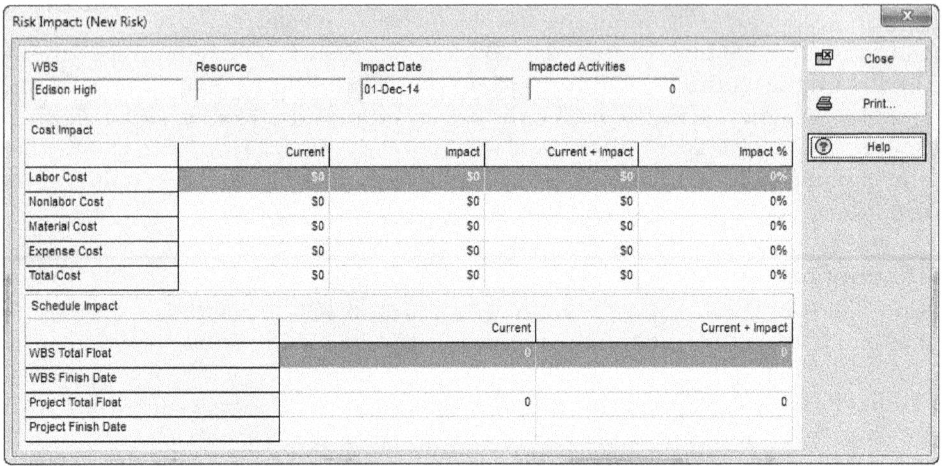

Figure 6-3 The **Risk Impact: (New Risk)** *dialog box*

In this dialog box, you can view the risk impact on the open project in terms of schedule, cost, and duration. The WBS and the resource on which the risk impact will be calculated will be displayed in the **WBS** and **Resource** edit boxes, respectively. The **Impact Date** edit box will display the date on which the risk impact will be calculated and the number of activities impacted due to risk will be displayed in the **Impacted Activities** edit box.

In the **Cost Impact** area, you can view the current cost of labor, non labor and material cost and the impact of the risk caused on these units. In the **Current+Impact** column, the sum of actual cost and the risk impacted cost is displayed. The impact % is calculated as the {Impact / (Current + Impact)}* 100 and is displayed under the **Impact %** column.

In the **Schedule Impact** area, the **Current** column displays the actual values that are assigned in the project for WBS Total Float, WBS Finish Date, Project Total Float, and Project Finish Date. The **Current + Impact** column displays the values changed due to the risk impact on a project.

THRESHOLDS

Threshold is generated in a project when certain values exceed the defined actual limit. If the threshold value exceeds, during the completion of a project, different issues might arise which will be handled by the responsible manager later on. The threshold is assigned to the WBS elements at the activities or at the WBS level.

You can view and create your own thresholds and assign them to the WBS elements to be monitored. To create thresholds, choose the **Thresholds** option from the **Project** menu; the **Project Thresholds** window will be displayed, as shown in Figure 6-4. Alternatively, you can choose the **Threshold** button from the Directory bar. Now, to add a threshold, choose the **Add** button from the Command bar; a new threshold will be added to the **Project Thresholds** window.

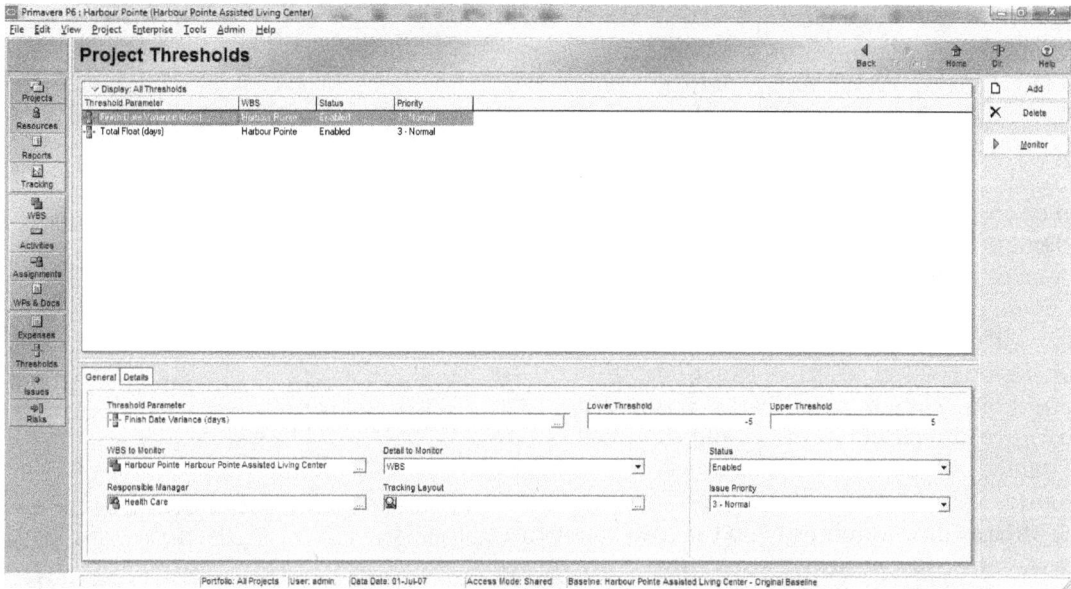

*Figure 6-4 The **Project Threshold** window with the **Details** table*

The **Project Thresholds** window displays the **Display** area and the **Threshold Details** table. In the **Display** drop-down, the **All Thresholds** option is selected by default. You can change the display by selecting the **Filter By** option from the **Display** drop-down. You can customize the column shown in the **Display** area by choosing the **Columns > Customize** option from the **Display** drop-down. According to the project, the thresholds are generated by default in the **Display** area.

Threshold Details Table

The **Threshold Details** table is displayed in the **Threshold** window and has two tabs, **General** and **Details**.

In the **General** tab, you need to enter the general information about the selected threshold and select the threshold parameter to which you want to assign the threshold values. To do so, choose the Browse button in the **Threshold Parameter** edit box; the **Select Threshold Parameter** dialog box will be displayed, as shown in Figure 6-5. In this dialog box, you can select the parameter to assign the threshold values as start date, finish date, total float, free float and so on and then choose the **Select** button. In the **Lower** and **Upper Threshold** edit boxes, you can enter the lower and upper threshold values. If any activity or WBS element values fall outside the threshold range, an issue will be generated.

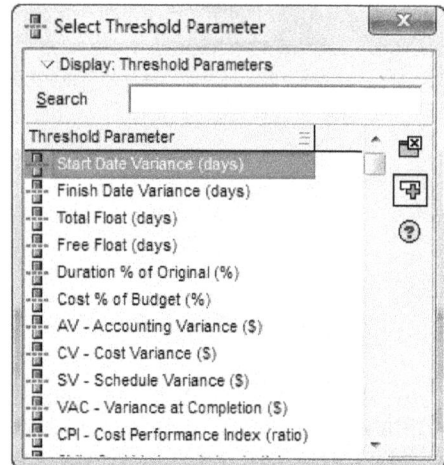

*Figure 6-5 The **Select Threshold Parameter** dialog box*

Select the **Activity** or **WBS** option from the **Detail to Monitor** drop-down list of the **General** tab. Then, select the required WBS element to be monitored on the basis of the assigned threshold values from the **WBS to Monitor** edit box. The manager responsible for addressing the issues generated by the threshold get automatically assigned when you select a WBS element. To assign a tracking layout to a threshold or an issue, choose the Browse button in the **Tracking Layout** edit box; the **Select Tracking Layout** dialog box will be displayed, as shown in Figure 6-6. In this dialog box, select the layout that best displays the threshold problem area. In the **Status** drop-down list, the threshold status is automatically set to **Enabled**. If you do not want to use it to monitor project, select the **Disabled** option from the **Status** drop-down list. You can also change the threshold issue priority by selecting the required option from the **Issue Priority** drop-down list.

Figure 6-6   The Select Tracking Layout dialog box

In the **Details** tab, you can view the issues generated by the assigned threshold values in the **Threshold Issues** area. In the **Monitor Time Window** area, assign the date to monitor the threshold for the WBS elements or activities. Using the Browse button, you can assign values in the **From Date** and **To Date** edit boxes. The module checks only those WBS elements or activities whose start dates are after the **From Date** and whose finish dates are before the **To Date** assigned.

Monitoring Thresholds

Once the threshold parameter is defined, you need to monitor the threshold to generate any applicable issues. Before monitoring the threshold, choose the **Monitor Thresholds** option from the **Tools** menu; the **Monitor Threshold** dialog box will be displayed, as shown in Figure 6-7.

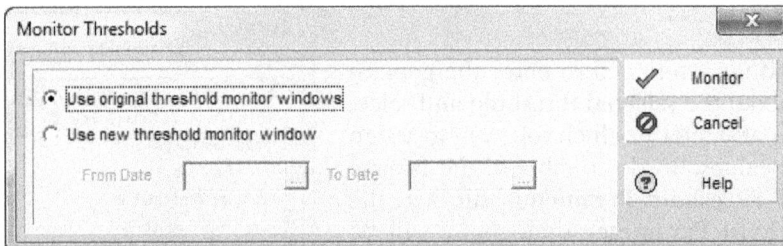

Figure 6-7 The Monitor Thresholds dialog box

In this dialog box, the **Use original threshold monitor windows** radio button allows you to monitor the threshold settings specified in the **General** tab by choosing the **Monitor** button in the **Project Threshold** window. The **Use new threshold monitor window** radio button allows you to specify the start date and end date of monitoring the threshold in the **From Date** edit box and the **To Date** edit box, respectively. Choose the **Monitor** button in the **Monitor Thresholds** dialog box; the **Primavera P6** message box will be displayed prompting you to monitor the selected threshold. Choose the **OK** button; the threshold will be monitored and the number of issues generated will be displayed in the **Primavera P6** message box, as shown in Figure 6-8. Choose the **Yes** button; the details of the issues will be displayed in the **Threshold Issues** area of the **Details** tab, as shown in Figure 6-9.

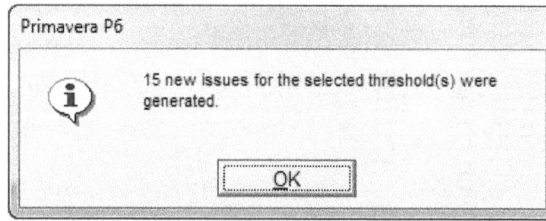

*Figure 6-8 The **Primavera P6** message box*

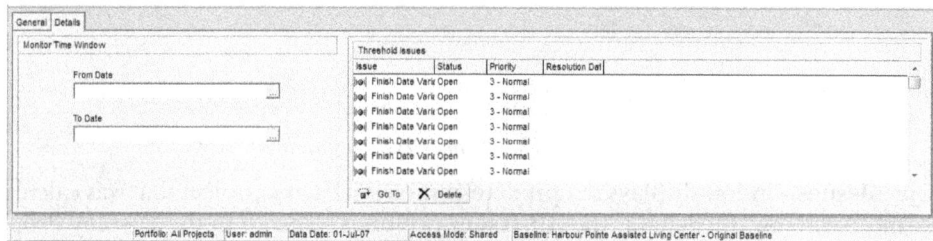

*Figure 6-9 The **Details** tab with the threshold issues*

Alternatively, you can choose the **Monitor** button from the Command bar; the **Primavera P6** dialog box will be displayed. Choose the **Yes** button if you want to monitor the threshold. On doing so, the monitoring starts and the **Primavera P6** message box is displayed with the number of issues generated.

ISSUES

Issues are automatically created by thresholds or they can be created manually. They are used to document, track, and monitor the arising problems and they may not have any impact on the cost or the schedule. An issue is different from the risk in a way that the risk is a preassumed condition that has not yet occurred in a project whereas an issue is a former risk that had already occurred in the project.

Issues are of two types: Issues originated from the conditions defined in the threshold and that are displayed in the **Threshold Issues** area. The other type of issues are originated when certain conditions are assumed.

To create an issue, choose the **Issues** button from the Directory bar; the **Project Issues** window will be displayed, as shown in Figure 6-10. Alternatively, you can choose the **Issues** option from the **Project** menu to display this window.

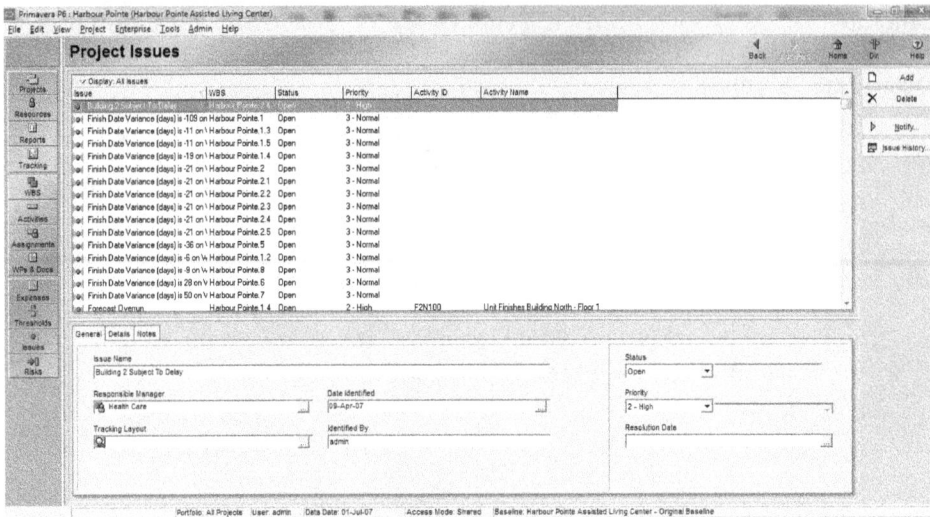

Figure 6-10 The Project Issues window

The **Project Issues** window displays the predetermined issues in a project that was calculated in the **Threshold** window. By default, the **All Issues** option is selected in the **Display** drop-down list. Below the **Project Issues** window, the **Issues Details** table is displayed. In the **Project Issues** window, you can add your own issue by choosing the **Add** button from the Command bar. On doing so, an issue will be assigned with the **New Issue** name in the **Project Issues** window.

To assign the details to the added issue, you need to use the **Issues Details** table. The table is described in detail next.

Issues Details Table

The **Issues Details** table consists of three tabs: **General**, **Details**, and **Notes**. In the **General** tab, you can enter the general information about an issue. Enter the name of the issue in the **Issue Name** edit box. Choose the Browse button in the **Responsible Manager** edit box and select the manager from the list displayed to address the issue. The date on which the issues is to be identified is automatically assigned in the **Date Identified** edit box. Choose the Browse button in the **Date Identified** edit box if you need to enter a different date. Choose the Browse button in the **Tracking Layout** edit box and select the layout from list displayed to assign the layout to an issue. Assign the status to the issue by selecting the **Open**, **Closed**, or **On Hold** option from the **Status** drop-down list. In the **Resolution Date** area, enter the date by which an issue will be resolved depending on the priority of the issue.

The details of the issues are to be entered in the **Details** tab. Assign value to the issue in the **Actual Value** edit box. To associate the issue with a WBS element other than the root WBS element, choose the Browse button in the **Applies to WBS** edit box; the **Select WBS** dialog box will be displayed. In this dialog box, select the WBS element to which you want to associate the issue.

In the **Notes** tab, you can enter additional information about the issues. You can use HTML editing features such as formatting text, inserting pictures, copying and pasting information from other document files and adding hyperlinks to enter additional information.

Issue History

You can view and add comments about the history of some selected issues. To add the history to an issue, choose the **Issue History** button from the Command bar; the **Issue History** dialog box will be displayed, as shown in Figure 6-11. In this dialog box, enter the history about the issue in the **Add to Notes** area. Now, choose the **Add** button; the information given in the **Add to Notes** area will be added in the **Issue History Notes** and will be assigned as the history of the selected issue, refer to Figure 6-11.

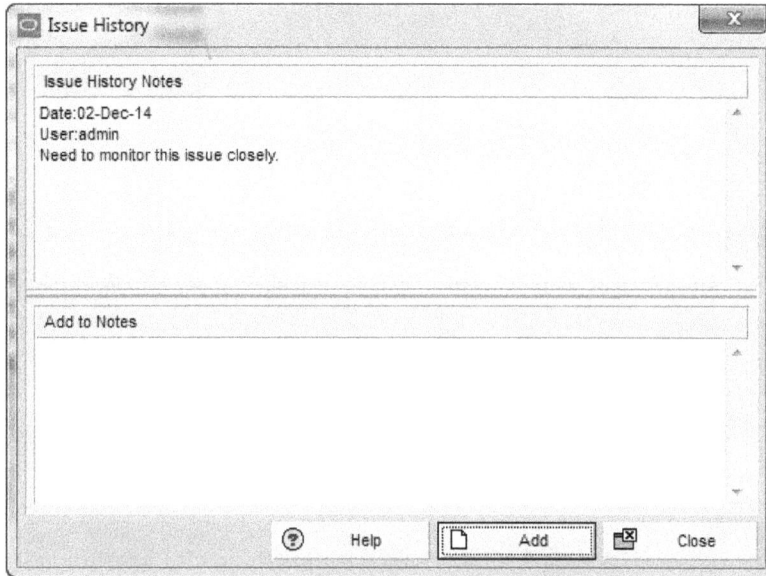

*Figure 6-11 The **Issues History** dialog box*

BASELINE

Baseline of a project is the reference used to compare or check the progress of the project at any given stage.

To create the baseline, two operations are required: **Managing Baselines** and **Assigning Baselines**. Baselines are created at one place and are assigned to the project for comparison at another place. It is done because in a project, different roles and resources perform different functions. As the baseline is used to calculate the variances, it would be risky if you create it on your own. Therefore, you will maintain the baseline from the existing project. Now, the maintained baseline will be assigned to the selected project.

Managing Baselines

To manage a baseline, you need to open the projects for which you want to create a baseline or view assigned baseline projects. To do so, choose the **Maintain Baselines** option from the **Project** menu; the **Maintain Baselines** dialog box will be displayed, as shown in Figure 6-12. In this dialog box, the open projects will show the preassigned baselines, if any, refer to Figure 6-12. In Figure 6-12, the **Edison Area High School** project has three baselines under it. First of them is the original and other two are updated. If a project has no preexisting baselines, you need to add a baseline under that project.

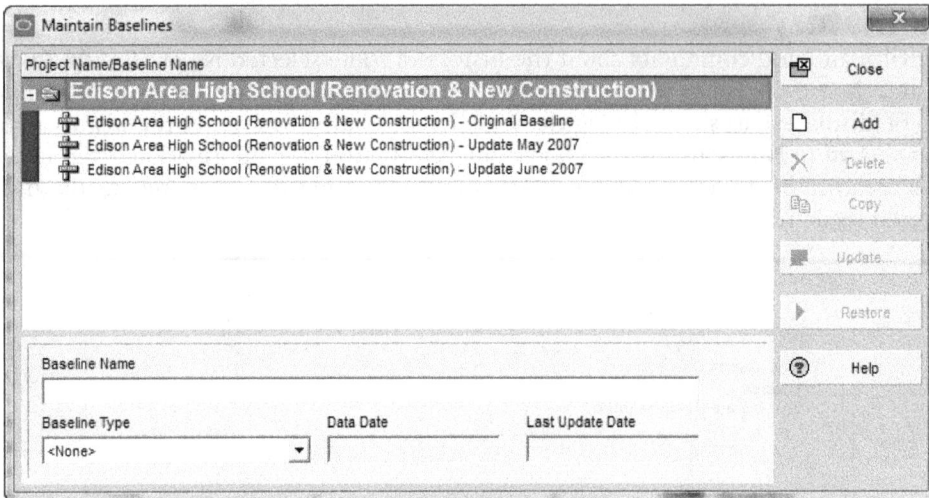

*Figure 6-12 The **Maintain Baselines** dialog box with the baselines*

Adding a Baseline

To add a baseline to a project, choose the **Add** button from the Command bar; the **Add New Baseline** dialog box will be displayed, as shown in Figure 6-13.

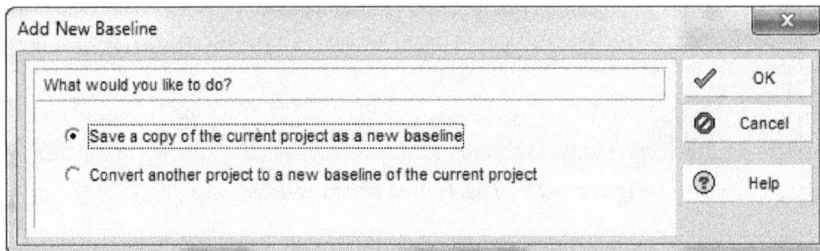

*Figure 6-13 The **Add New Baselines** dialog box*

This dialog box contains two radio buttons. If you select the **Save a copy of the current project as a new baseline** radio button, then a new baseline will be created which will be a copy of the current project. The **Convert another project to a new baseline of the current project** radio button allows you to choose a different project to use as a baseline for the selected project. On doing so, the project will get listed in the **Projects** window.

Note that you cannot select a currently opened project as the baseline, nor you can select a project that already has its own baseline assigned. Select the required radio button from the **Add New Baseline** dialog box and then choose the **OK** button; the **Select Project** dialog box will be displayed, as shown in Figure 6-14. In this dialog box, select the desired project and choose the **Select** button; the selected project will be assigned as baseline to the current project.

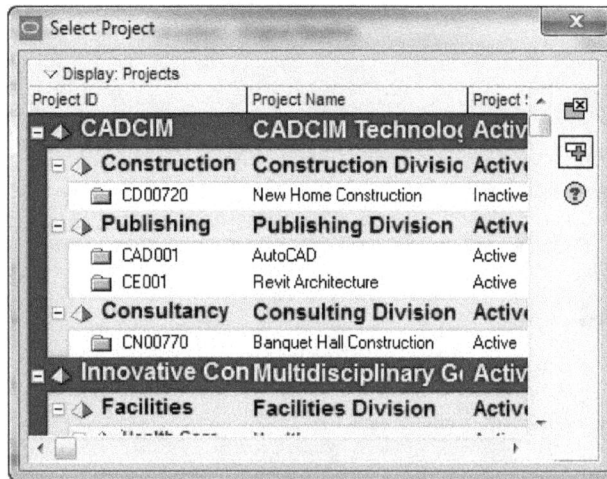

Figure 6-14 *The* **Select Project** *dialog box*

Deleting a Baseline

You can delete a baseline from the project database. To do so, open the project which contains the baseline and select the **Maintain Baselines** option from the **Projects** menu of the menubar; the **Maintain Baseline** dialog box will be displayed. In the **Maintain Baseline** dialog box, select the project that you want to delete and then choose the **Delete** button from the Command bar.

Note

You cannot delete any active baseline which is already assigned to the project.

Updating a Baseline

While working on an original baseline project, sometimes the information in the baseline needs to be updated to reflect the changes in the project. To update the baseline, you need to open the **Maintain Baselines** dialog box and then choose the baseline to be updated. Now, choose the **Update** button from the Command bar of the **Maintain Baseline** dialog box; the **Update Baseline** dialog box will be displayed, as shown in Figure 6-15.

In this dialog box, the **When updating project data, include** area allows to determine the entities to be updated. You can select any one check box from the area to update the information in the baseline. The **Specify the activities to include** area allows you to update the activities. On selecting the **All activities** radio button in this area, you can update all the activities. The **Activities within the following filter** radio button allows you to update those activities that match a filter. Select the **Activities within the following filter** radio button, an edit box under the radio button will be enabled. Choose the Browse button from that edit box; the **Filters** dialog box will be displayed, as shown in Figure 6-16.

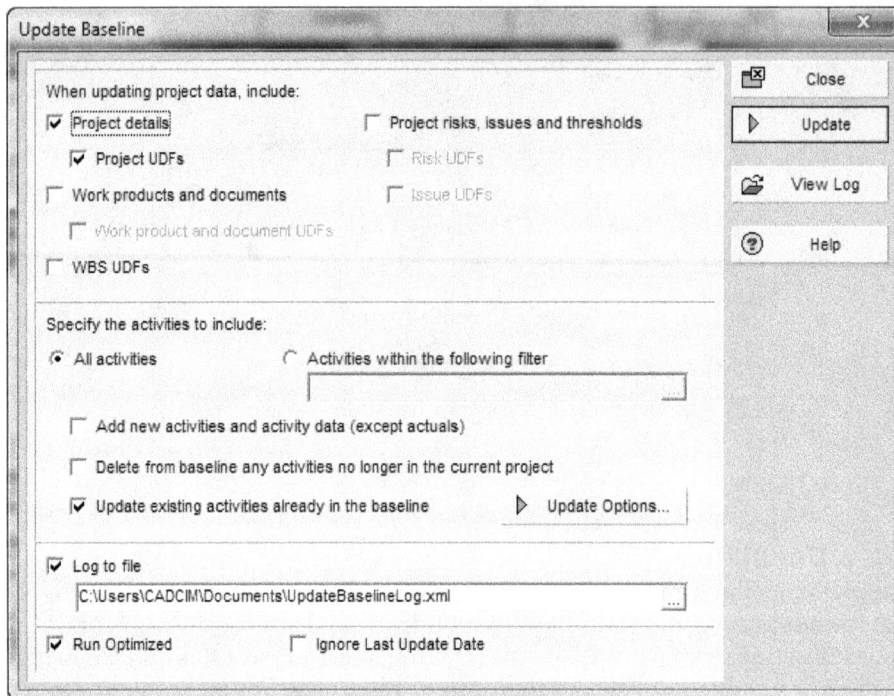

Figure 6-15 *The **Update Baseline** dialog box*

The **Filters** dialog box allows you to choose any activity from the set of pre-defined filters, or you can create your own filter. In this dialog box, the **All Activities** check box is selected by default. You cannot clear this check box until any activity type is selected from the **Filter** area. You can choose to update activities which are in the project but are not in the baseline. After selecting the activity type, choose the **OK** button; the **Filters** dialog box will be closed and settings will be done according to the selected filter.

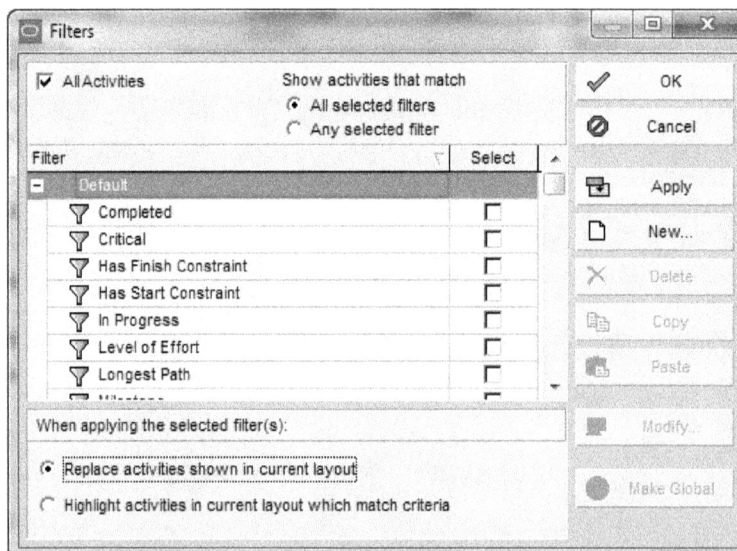

Figure 6-16 *The **Filters** dialog box*

Restoring a Baseline

To restore baseline of a project, open that project. Choose the **Restore** button; the **Primavera P6** message box will be displayed, as shown in Figure 6-17. Choose the **Yes** button to unlink the selected baseline and make it an individual project. The restored project is placed as project in the same node to which it was linked as baseline.

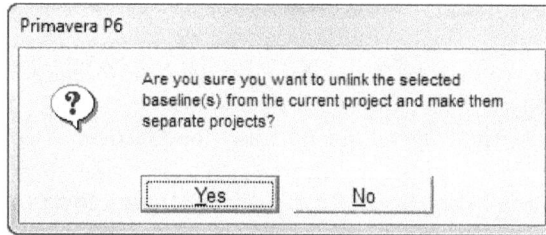

*Figure 6-17 The **Primavera P6** message box*

Assigning Baseline

In the previous section, you learned to create and maintain a baseline. Now, you need to assign the baseline so that project analysis can be done according to the baseline. There are two types of baseline assignments: **Project** and **User**. To assign baseline to a project, select the **Assign Baseline** option from the **Project** menu; the **Assign Baselines** dialog box will be displayed, as shown in Figure 6-18.

In this dialog box, the name of the project will be displayed in the **Project** drop-down list by default. In the **Project Baseline** drop-down list, you can select the baseline that is needed to be assigned in the project. In the **User Baselines** area, you can select the baseline from the **Primary** drop-down list, and if needed from the **Secondary** and **Tertiary** drop-down lists as well. After making the required settings, choose the **OK** button; the baseline will be assigned to the project.

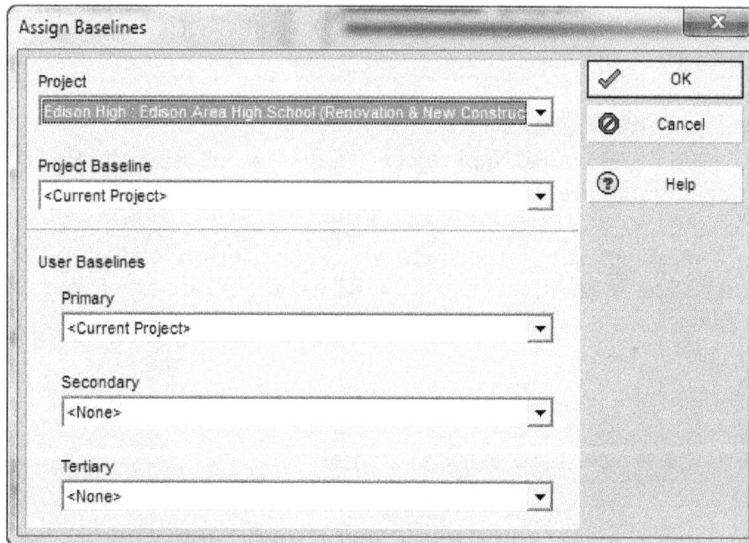

*Figure 6-18 The **Assign Baselines** dialog box*

TUTORIALS

To perform the tutorials of this chapter, you need to complete Tutorial 1 and Tutorial 2 of Chapter 2.

To perform this tutorial, you need *c05_Primavera_P6_tut.zip* file. You can download this file from *www.cadcim.com* using the following steps:

1. To download the file, browse to *Textbooks > Civil/GIS> Primavera P6 > Exploring Oracle Primavera P6 v7.0*. Next, select the *c05_Primavera_P6_tut.zip* file from the **Tutorial Files** drop-down list. Choose the corresponding **Download** button to download the data file.

2. Now, save and extract the downloaded file to the following location:

    C:/PM6

3. Import the *c05_CONS_Home_tut02* file from the extracted folder to the Construction EPS created in Tutorial 1 of Chapter 2.

Tutorial 1   Risks, Thresholds & Issues

In this tutorial, you will assign risk and thresholds to the **Home Construction** project. On assigning the thresholds, the issues will be calculated automatically.

(Expected time: 50 min)

The following steps are required to complete this tutorial:

a. Open the **Project Risks** window.
b. Add a risk to the project.
c. Add another risk to the project.
d. Assign threshold to the risk.
e. Export the project.

Opening the Project Risks Window

1. Open Primavera P6 and invoke the **Projects** window by choosing the **Projects** option from the **Enterprise** menu in the menubar.

2. In this window, select the project **Home Construction** project placed under the **Construction** EPS node and then choose the **Risks** button from the Directory bar to display the **Project Risks** window.

Adding Risks

1. In this window, choose the **Add** button from the Command bar; the risk with the name **New Risk** is displayed.

2. Ensure that the **Risks Details** table is displayed and choose the **General** tab in the **Risks Details** table.

3. Enter the text **Non availability of masonry crew** in the **Risk Name** edit box of the **General** tab.

4. Choose the Browse button in the **Applies to WBS** edit box; the **Select WBS** dialog box is displayed.

5. Choose the **Home Construction** option from the **Select WBS** dialog box.

6. Choose the **Select** button; the WBS is assigned and the dialog box is closed.

7. Choose the Browse button in the **Applies to Resource** edit box; the **Select Resource** dialog box is displayed, as shown in Figure 6-19.

8. In this dialog box, select the **Masonry Crew** option from the **Resource Name** column and then choose the **Select** button.

9. Choose the Browse button in the **Risk Type** edit box; the **Select Risk Type** dialog box is displayed, as shown in Figure 6-20.

*Figure 6-19 The **Select Resource** dialog box*

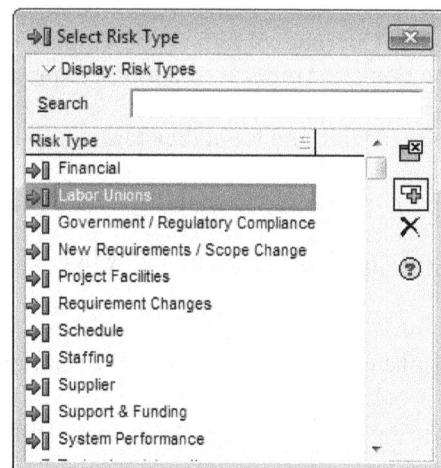

*Figure 6-20 The **Select Risk Type** dialog box*

10. In this dialog box, select the **Labor Unions** option under the **Risk Type** column and then choose the **Select** button.

11. Select the **2-High** option from the **Priority** drop-down list in the **Risk Details** table.

12. In the **Risk Details** table, choose the Browse button in the **Date Identified** edit box and then assign the date **13-Oct-15**.

13. Choose the **Impact** tab and then assign the date **13-Oct-15** in the **Impact Date** edit box.

   After the **Impact Date** is assigned, the **Impact Tab** displays complete information about the risk, as shown in Figure 6-21. The **Impacted Activities** edit box displays number of activities to be impacted as **4**, refer to Figure 6-21, indicating that 4 activities are being affected due to non availability of masonry crew.

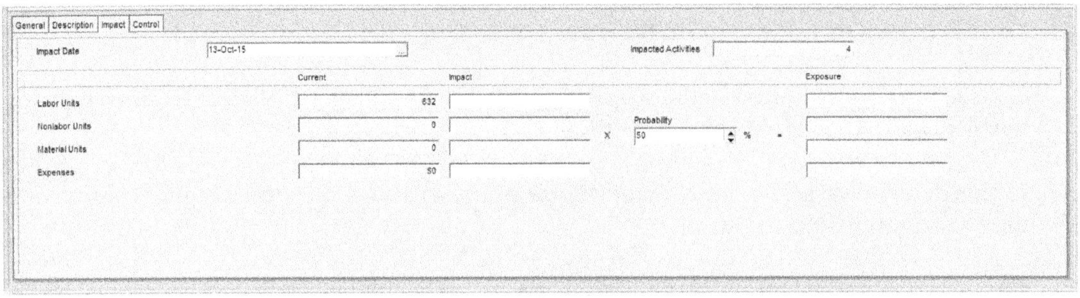

*Figure 6-21 The **Impact** tab with the detailed information of risk*

   As the assigned resource is a nonlabor unit, therefore the current non labor units are assigned in the **Current** column of the **Impact** tab. The unavailability of masonry crew will impact the nonlabor units as well as the expenses.

14. Enter **150** in the **Nonlabor Units** edit box under the **Impact** column.

15. Enter **75%** in the **Probability** edit box as the chances for the impact to occur are 75%.

16. Enter the impact expenses as **$20** in the **Expenses** edit box under the **Impact** column.

17. Choose the **Calc Impact** button from the Command bar; the **Risk Impact** dialog box is displayed with the complete information of the risk, as shown in Figure 6-22.

18. Choose the **Close** button; the **Risk Impact** dialog box is closed.

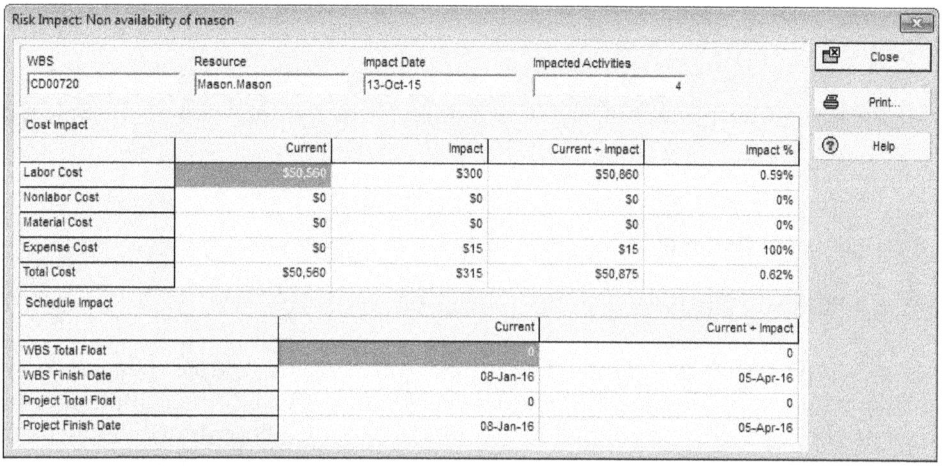

*Figure 6-22 The **Risk Impact** dialog box with the calculated impact values*

Adding Another Risk

1. Choose the **Add** button from the Command bar; the new risk is added to the **Project Risks** window.

2. Ensure that the **General** tab is chosen in the **Risks Details** table.

3. Enter the text **Non availability of roof sheeting manpower** in the **Risk Name** edit box.

4. Double-click under the **WBS** column in the **Risks** window; the **Select WBS** dialog box is displayed.

5. In this dialog box, select the **Roof** option from the **WBS Name** column and then choose the **Select** button.

6. Choose the Browse button in the **Applies to Resource** edit box; the **Select Resource** dialog box is displayed.

7. In this dialog box, select the **Skilled Laborer** option from the **Resources Name** column and then choose the **Select** button.

8. Choose the Browse button in the **Responsible Manager** edit box; the **Select Responsible Manager** dialog box is displayed.

9. In this dialog box, select the **Project Manager** option from the **OBS Name** column and then choose the **Select** button.

10. Choose the Browse button in the **Risk Type** edit box; the **Select Risk Type** dialog box is displayed.

11. Select the **Schedule** option from the **Risk Type** column and then choose the **Select** button.

12. Select the **1-Top** option from the **Priority** drop-down list.

13. In the **Date Identified** edit box, assign **18-Dec-15** as the identified date.

14. Choose the **Impact** tab and assign the **Impact Date** as **19-Dec-15**.

    After the **Impact Date** is assigned, the **Impact Tab** displays complete information about the risk, as shown in Figure 6-23. The **Impacted Activities** edit box displays number of activities to be impacted as **2** from 19-Dec-15 due to unavailability of roof sheeting manpower.

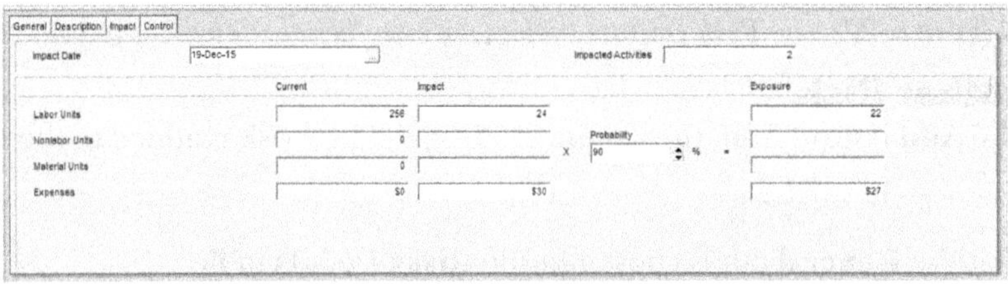

Figure 6-23 The Impact tab with the detailed information of another risk

15. Enter **3** in the **Labor** units edit box under the **Impact** column.

16. Enter **90%** in the **Probability** edit box as the chances for this impact to occur are 90%.

17. Enter the impact expenses as **$30** in the **Expenses** edit box under the **Impact** column.

18. Choose the **Calc Impact** button from the Command bar; the **Risk Impact** wizard is displayed with complete information of the risk, as shown in Figure 6-24.

19. Choose the **Close** button; the wizard is closed.

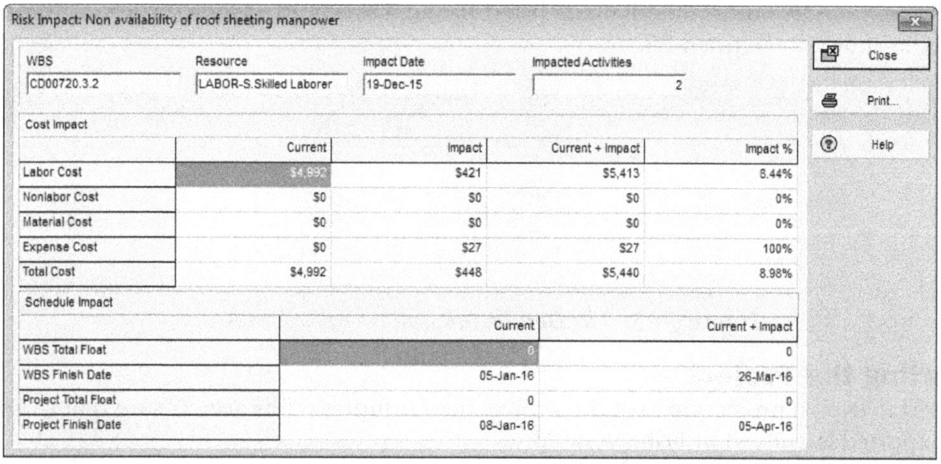

*Figure 6-24   The **Risk Impact** dialog box with the calculated impact values*

Assigning Thresholds

1. Choose the **Thresholds** button from the Directory bar; the **Project Thresholds** window is displayed.

2. Choose the **Add** button from the Command bar; the threshold parameter gets added.

3. Choose the Browse button in the **Threshold Parameter** edit box in the **General** tab; the **Select Threshold Parameter** dialog box is displayed.

4. In this dialog box, select the **Start Date Variance (days)** option from the **Threshold Parameter** column and then choose the **Select** button.

5. Choose the Browse button in the **WBS to Monitor** edit box; the **Select WBS** dialog box is displayed.

6. In this dialog box, choose the **Home Construction** option from the **WBS** column and then choose the **Select** button.

7. Choose the Browse button in the **Responsible Manager** edit box; the **Select Responsible Manager** dialog box is displayed.

8. Choose the **Project Manager** option from the dialog box and then choose the **Select** button.

9. Select the **WBS** option from the **Detail to Monitor** drop-down list.

10. Enter **0** in the **Lower Threshold** edit box and **2** in the **Upper Threshold** edit box.

11. Choose the **Monitor** button from the Command bar; the **Primavera P6** message box is displayed.

12. Choose the **Yes** button; another **Primavera P6** dialog box is displayed with the number of issues generated.

13. Choose the **OK** button; the issue is listed in the **Threshold Issues** section of the **Details** tab, as shown in Figure 6-25.

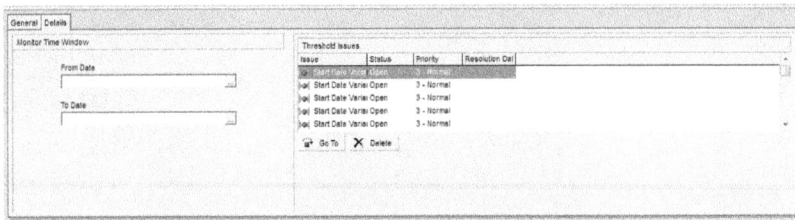

Figure 6-25 *The **Details tab** with the calculated issues*

Exporting the Project

You need to export projects to save the project file for future reference. Ensure that the project to be exported is opened in Primavera P6.

1. Choose the **Export** option from the **File** menu; the **Export** wizard with the **Export Format** page is displayed.

2. In this page, ensure that the **Primavera PM/MM - (XER)** radio button is selected. Next, then choose the **Next** button; the **Export Type** page is displayed.

3. In this page, ensure that the **Project** radio button is selected and then choose the **Next** button; the **Projects To Export** page with the **Home Construction** project is displayed.

4. Choose the **Next** button; the **File Name** page is displayed.

5. In this page, choose the Browse button from the **File Name** edit box; the **Save File** dialog box is displayed.

6. In this dialog box, browse to *C:\PM6* and then create a sub-folder with the name **c06**.

7. Open the created folder and save the file with the name **c06_CONS_HOME_tut01**. Then, choose the **Save** button.

8. Choose the **Finish** button; the **Primavera P6** message box is displayed with the message that the export was successful.

9. Choose the **OK** button; the file gets exported.

Tutorial 2   Setting Baseline

In this tutorial, you will create the baseline and then assign it to the **Home Construction** project.
                                                                    (Expected time: 40 min)

The following steps are required to complete this tutorial:

a. Open the **Projects** window.
b. Maintain the baseline.

c.   Assign the baseline.

d.   Export the project.

Opening the Projects Window

1.   Open Primavera P6 and invoke the **Projects** window by choosing the **Projects** option from the **Enterprise** menu in the menubar.

2.   In this window, select the **Home Construction** project placed under the **Construction** node.

Maintaining the Baseline

1.   Choose the **Maintain Baselines** option from the **Project** menu; the **Maintain Baselines** dialog box is displayed, as shown in Figure 6-26.

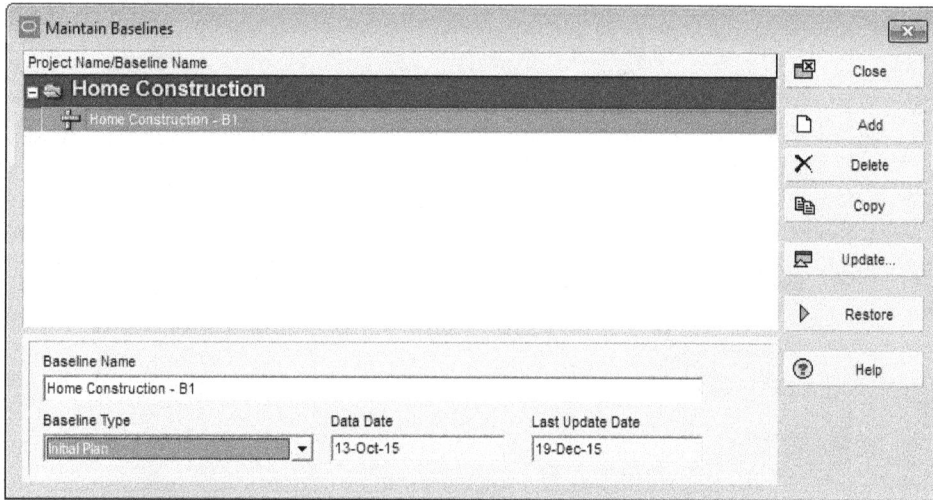

*Figure 6-26   The **Maintain Baselines** dialog box with the baseline*

2.   In this dialog box, choose the **Add** button from the Command bar; the **Add New Baseline** dialog box is displayed, as shown in Figure 6-27.

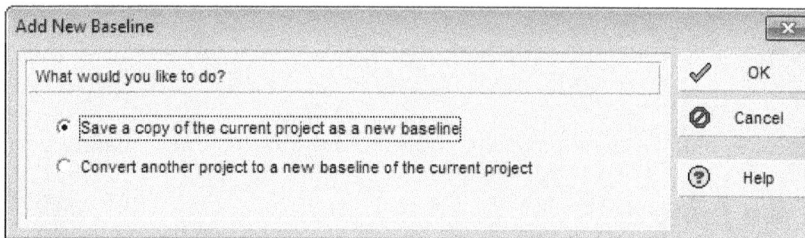

*Figure 6-27   The **Add New Baseline** dialog box*

3.   In this dialog box, ensure that the **Save a copy of the current project as a new baseline** radio button is selected.

4.   Choose the **OK** button; a new baseline with the name **Home Construction - B1** is added under the **Home Construction** head.

5. In the **Maintain Baseline** dialog box, select the **Initial Plan** option from the **Baseline Type** drop-down list.

6. Ensure that the data date and last update date is entered in the **Data Date** and **Last Update Date** edit boxes, refer to Figure 6-26.

7. Choose the **Close** button; the dialog box is closed.

Assigning the Baseline

1. Choose the **Assign Baselines** option from the **Project** menu; the **Assign Baselines** dialog box is displayed, as shown in Figure 6-28.

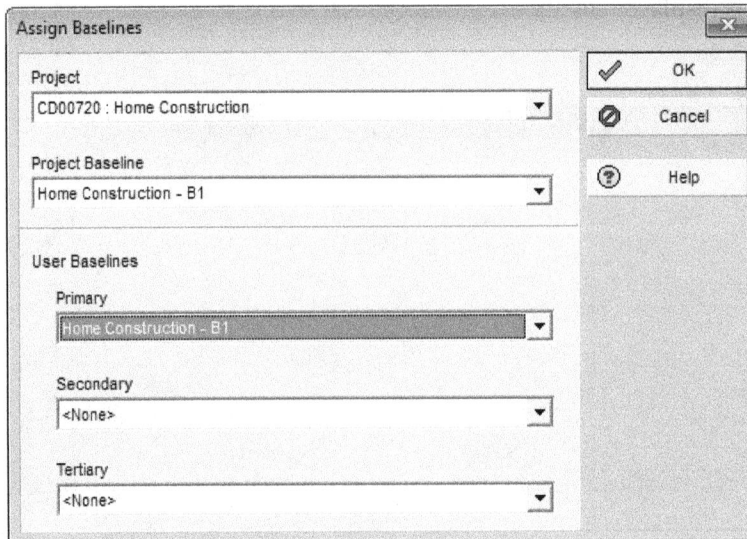

*Figure 6-28   The **Assign Baselines** dialog box with the assigned baseline*

2. In this dialog box, select the **Home Construction - B1** option from the **Project Baseline** drop-down list, refer to Figure 6-28.

3. Similarly, select the **Home Construction - B1** option from the **Primary** drop-down list, refer to Figure 6-28.

4. Choose the **OK** button; the dialog box is closed and the baseline is assigned to the project.

Exporting the Project

You need to export projects to save the project file for future reference. Ensure that the project to be exported is opened in Primavera P6.

1. To export a project, choose the **Export** option from the **File** menu; the **Export** wizard with the **Export Format** page is displayed.

2. In this page, ensure that the **Primavera PM/MM - (XER)** radio button is selected and then choose the **Next** button; the **Export Type** page is displayed.

3. In this page, ensure that the **Project** radio button is selected and then choose the **Next** button; the **Projects To Export** page with the **Home Construction** project is displayed.

4. Choose the **Next** button; the **File Name** page is displayed.

5. In this page, choose the Browse button next to the **File Name** edit box; the **Save File** dialog box is displayed.

6. In this dialog box, browse to the *C:\PM6\c06*.

7. Save the file with the name **c06_CONS_HOME_tut02**. Then, choose the **Save** button.

8. Choose the **Finish** button; the **Primavera P6** message box is displayed with the message that the export was successful.

9. Choose the **OK** button; the file gets exported.

Self-Evaluation Test

Answer the following questions and then compare them to those given at the end of this chapter:

1. Which of the following options enables you to invoke the **Project Risks** window?

   (a) **View**   (b) **Project**
   (c) **Enterprise**   (d) **Tools**

2. Which of the following windows allows you to calculate impact?

   (a) **Project Issues**   (b) **Project Thresholds**
   (c) **Project Risks**   (d) **Project Expenses**

3. Which of the following tabs allows you to calculate the risk and the exposure values?

   (a) **Impact**   (b) **Description**
   (c) **General**   (d) **Control**

4. Which of the following conditions affects the impact date?

   (a) Risks   (b) Threshold
   (c) Issues   (d) Facilities

5. Which of the following affects the labor and non labor units?

      (a) **At Completion Units**   (b) **Budgeted Units**
      (c) **Max/Min Units**   (d) **Remaining Units**

6. Issues are generated only when the thresholds are assigned. (T/F)

7. Risks can affect the cost involved in a project. (T/F)

8. You cannot create your own baseline. (T/F)

9. You cannot calculate the risk impact without assigning the deadline date. (T/F)

10. You can monitor thresholds by using the **Monitor** button. (T/F)

Review Questions

Answer the following questions:

1. The _____ tab helps you to write about the control plan.

2. The _____ dialog box allows you to assign baseline to a project.

3. The _____ tab of the **Project Threshold** window allows to view the issues.

4. _____,_____,_____ are the resource types.

5. The _____ edit box allows you to assign the type of risk to a project.

6. It is necessary to assign the baseline in the **Secondary** and **Tertiary** drop-down list. (T/F)

7. You cannot assign more than one threshold from the **Project Threshold** window. (T/F)

8. You can monitor the threshold by assigning the dates. (T/F)

9. You can add issue history to the **Project Issues** window. (T/F)

10. You cannot update a baseline once it is created. (T/F)

EXERCISES

Exercise 1

In this exercise, you will add risks and assign threshold to the **Hospital Building** project.

(Expected time: 40 min)

Hints:
1.   Add the risk as non availability of labor and assume other values.
2.   Assign the threshold according to the risk impact and assume other values.

Exercise 2

In this exercise, you will create baseline and then assign it to the **Hospital Building** project.

(Expected time: 45 min)

Hints:
1.   Maintain the baseline as the project of baseline.
2.   Assign it to the project.

Answers to Self-Evaluation Test

1. b, 2. c, 3. a, 4. c, 5. b, 6. T, 7. T, 8. F, 9. F, 10. T

Chapter 7

Project Expenses and Tracking Progress of Project

Learning Objectives

After completing this chapter, you will be able to:
- *Create cost accounts*
- *Calculate expenses*
- *Use the expenses details table*
- *Update the project status*
- *Track the progress of the project*
- *Group, sort, and filter data*

INTRODUCTION

In the previous chapter, you learned to add and manage risks assign thresholds and risks to generate issues and assign baseline to keep record of project.

In this chapter, you will learn to calculate the expenses of a project. Expenses are calculated to make the project cost effective. After the calculation of expenses, the project status can be updated every week. By updating the project status, you can track the project delays. In this chapter, you will learn how to track the progress of project and how to create and maintain tracking layout. Moreover, you will learn how to customize the format and level of information that each tracking layout displays.

COST ACCOUNTS

Cost accounts help you to track the activity costs and also the earned value of the overall project. You need to set the cost account at the project level and will be automatically assigned to the project activities. You can set up a cost account structure and can assign activity codes. The cost account structure helps to track the amount of work accomplished against the amount of money spent on work.

Creating Cost Accounts

You can create your own cost account structure. To do so, choose the **Cost Accounts** option from the **Enterprise** menu in the menubar; the **Cost Accounts** dialog box will be displayed, as shown in Figure 7-1. This dialog box displays two columns: **Cost Account ID** and **Cost Account Name**. The **Cost Account ID** column displays the ID of the cost accounts in the hierarchial structure. In this dialog box, by default the **Current Project's Cost Account** option is selected in the **Display** drop-down. To assign the cost accounts to the project, choose the **All Cost Accounts** option from the cascading menu displayed on choosing the **Filter By** option in the **Display** drop-down, refer to Figure 7-1. You can add cost account above or at the same level by choosing the **Add** button from the Command bar. On doing so, a row will be added in the **Cost Accounts** dialog box and then you can enter the id in the **Cost Account ID** column and name in the **Cost Account Name** column. Add description about the cost account added in the **Cost Account Description** area .

Assigning the Cost Account

It is recommended to use the default cost account that will be used for the resources assigned to activities and project expenses in the selected project. To assign the cost account to the project, select the project from the **Projects** window and choose the **Defaults** tab from the **Projects Details** table. In this tab, click on the Browse button corresponding to the **Cost Account** parameter; the **Select Default Cost Account** dialog box will be displayed, as shown in Figure 7-2. You can change the filter of this dialog box by choosing the **All Cost Accounts** option from **Display > Filter By** flyout; all the cost accounts will be displayed, refer to Figure 7-2. You can select the required cost account from the list and then choose the **Select** button; the required cost account will be assigned to the project and the dialog box will be closed.

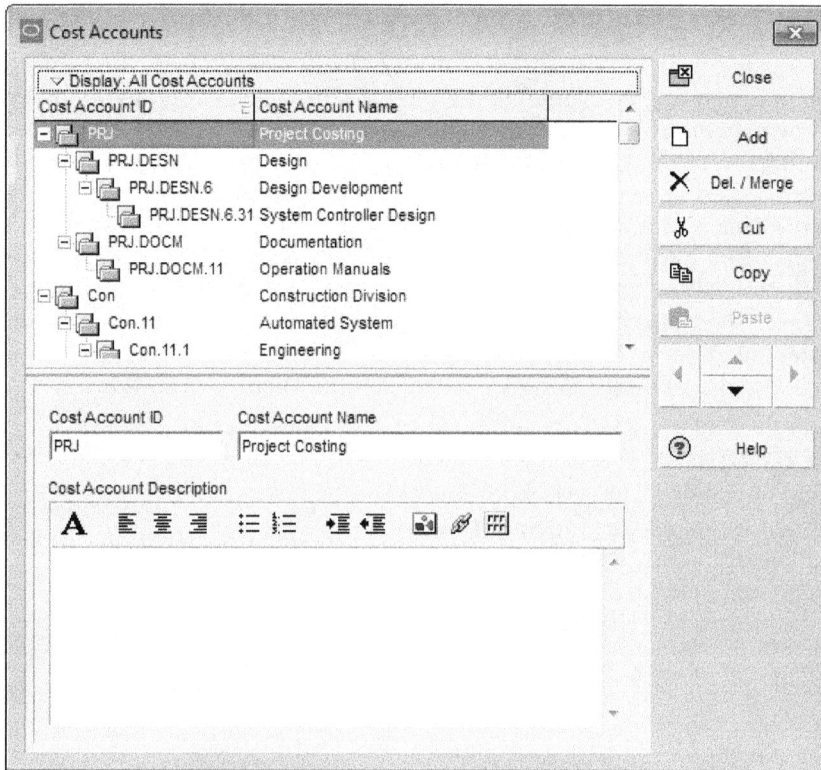

*Figure 7-1 The **Cost Accounts** dialog box*

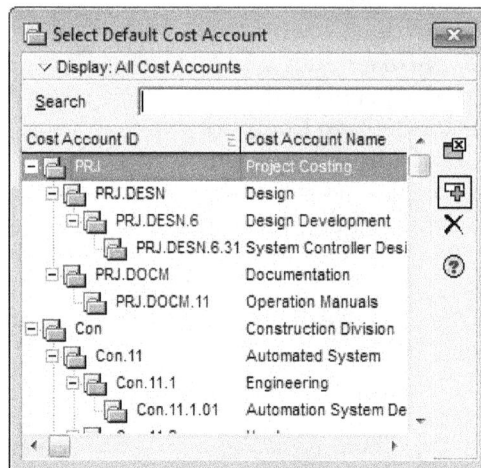

*Figure 7-2 The **Select Default Cost Account**
dialog box with all the cost accounts*

PROJECT EXPENSES AND COST INFORMATION

Expenses are related to the nonresource cost activities associated with the project and also if any such kind of expenses are assigned to the project's activities. Expenses are basically defined as the one time expenditure for non-reusable items. Examples of expenses include facilities, travel, consulting, and training.

To calculate the expenses and cost related information for the open project you can use the **Project Expense** window. In this window, you need to assign the cost account and the work breakdown structure (WBS) code so that the project component associated with the expense can be identified. The **Project Expenses** window is displayed on choosing the **Expenses** option from the **Project** menu. The **Expense Details** table is displayed by default in this window. If it is not, then choose the **Expense Details** option from the **Display** drop-down.

Adding Expenses

To add expenses incurred on a project, choose the **Add** button from the Command bar; the **Select Activity** dialog box will be displayed, as shown in Figure 7-3. In this dialog box, select the activity on which expense is to be incurred and then choose the **Select** button; the expense is added under the **No Expense Category** by default.

Now, select the expense for which the cost information is to be entered. Now, choose the **Costs** tab from the **Expense Details** table and enter the expected number of units for the expense's assigned activity and then supply the price for each unit. The Primavera will automatically calculate and display the budgeted cost of the selected expense (Budgeted Unit*Price/Unit). You can select the **Auto Compute Actuals** check box to automatically compute an

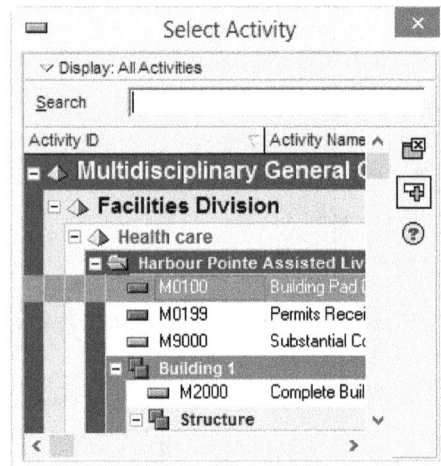

Figure 7-3 The Select Activity dialog box

expense's actual cost, based on the activity completion percentage. On selecting the **Auto Compute Actuals** check box, the remaining and actual values get calculated automatically. These settings in the module indicate that work is proceeding according to the plan.

You can enter other details in the **Activity** tab. Basically, in this tab, you need to enter the **Accural Type** for the expense required. There are three accural types: **Start of Activity**, **End of Activity**, and **Uniform over Activity**.

The **Start of Activity** option is used when the entire expense is accumulated on the start date of an activity.

The **End of Activity** option is used when the entire expense is accumulated on the date when the activity ends.

The **Uniform over Activity** option is used when the expenses are evenly distributed over the activity's duration.

Assigning Expense Category

Assigning an expense category enables you to classify the type of cost and also to group, sort, filter, and report the expense and the cost information of your project. To assign the expense category, choose the **General** tab from the **Expense Details** table and then choose the Browse button corresponding to the **Expense Category** parameter; the **Select Expense Category** dialog box will be displayed. This dialog box displays the list of expense categories. You can select the required expense category type from the list and then choose the **Select** button; the category will be assigned and dialog box will be closed.

EXPENSE DETAILS TABLE

The **Expense Details** table helps you to view the detailed information of the expenses incurred. Using this table, you can also edit the information related to expenses. The cost related information is also assigned in this table. This table is displayed by default in the **Project Expenses** window. You can hide or unhide the **Expense Details** table by choosing the **Expense Details** option from the **Display** drop-down.

The **Expense Details** table displays four tabs: **General**, **Activity**, **Costs**, and **Description**, refer to Figure 7-4. The description of these tabs is given next.

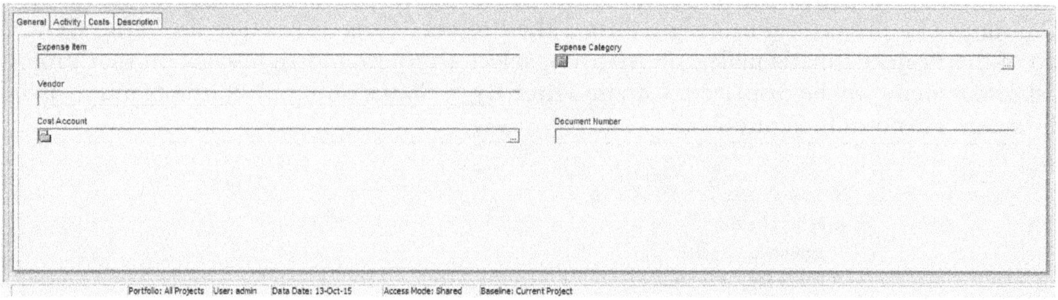

*Figure 7-4 The **Expense Details** table*

General Tab

In the **General** tab, you can provide the details of the expense items such as expense name, category, vendor name, cost account, and document number. To provide these details, you need to click on the field corresponding to the required parameter. Expense item and vendor name need to be assigned in the text format. The expense category and cost account can be assigned by clicking on the corresponding Browse button and then selecting the desired option from the dialog box displayed.

Activity Tab

The **Activity** tab is used to change the assigned activity for the created expense items. Other information is entered by default in this tab. You need to assign the accrual type for the expenses by selecting the required option from the **Accrual Type** drop-down list.

Costs Tab

The **Costs** tab is used to specify cost amounts for the selected expense items, including price/unit, budgeted cost, actual cost, and remaining cost. You can select the **Auto Compute Actuals** check box to compute the expenses automatically using the budgeted cost and activity's schedule percent complete.

Description Tab

The **Description** tab is used to write the description about the expense items. In this tab, you can type the new description related to expenses. To do so, the HTML editing features can be used which include formatting of text, inserting pictures, copying and pasting other information, and adding hyperlinks.

PROJECT STATUS

In Primavera P6, the regular updation of a project is necessary for keeping its track. The project status can be updated in the **General** tab of the **Project Details** table. The status of the project can be set as: **Planned**, **Active**, **What-if**, and **Inactive**. The **Project Status** field acts as a label by identifying the significance of the project when considering current planned work load. The **Planned** status is assigned to the project for which planning is going on for starting in future. The **Active** project status is assigned to the project that is live and is being executed. It should be regularly updated depending upon the progress. The **What-if** status defines that the project has the alternative scope of work. The **Inactive** project status is assigned to the project which is completed or is in the stage of completion when no more work is to be done on that project. The status of a project can be assigned from the **Project Status** drop-down list in the **General** tab of the **Project Details** table. Alternatively, select a project and right-click on that project; a shortcut menu will be displayed. Choose **Filter By > Status** option from the menu to assign the status, as shown in Figure 7-5.

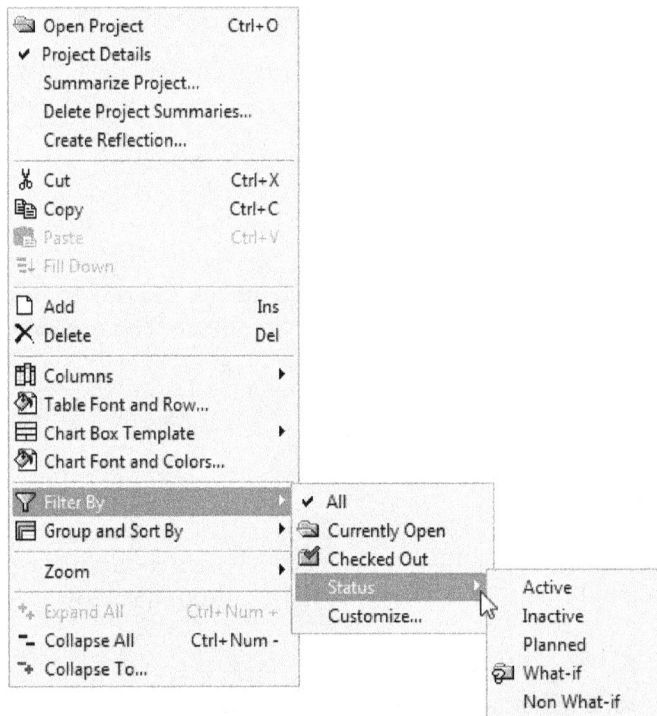

Figure 7-5 *Selecting the project status*

Updating Status

A project needs to be updated every week with the current data date. The progress spotlight enables you to view the progress of activities with the data date. To update the progress of an activity, choose the **Update Progress** option from the **Tools** menu; the **Update Progress** dialog box will be displayed, refer to Figure 7-6. You can update durations and actual values of the selected activities using the settings in this dialog box.

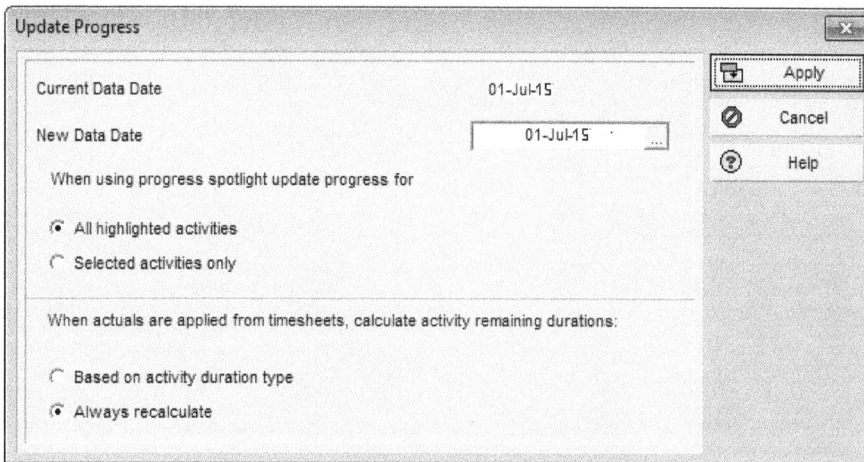

*Figure 7-6 The **Update Progress** dialog box*

In this dialog box, the current data date is entered by default. This is the starting date of the project. You can enter the data date according to the progress of the project by using the Browse button next to the **New Data Date** edit box. When you choose this button, a flyout will be displayed. Select the date from the calendar and then choose the **Select** button; the date will be assigned to the project and the progress will be updated according to the new data date. In this dialog box, you can select the **All highlighted activities** radio button in the **Update Progress** dialog box to update all the highlighted activities that appear in the Gantt Chart. To update the selected activities only, select the **Selected activities only** radio button. In the **When actuals are applied from timesheets, calculate activity remaining durations** area, select the **Based on activity duration type** radio button to recalculate the remaining duration of an activity based on the activity duration type. On selecting the **Always recalculate** radio button, the remaining duration will be calculated based on the fixed units or fixed units/time. Now, choose the **Apply** button; the dialog box will be closed and the progress will be updated and highlighted in yellow color.

TRACKING PROGRESS

Tracking is a process in which the status of a project is tracked. This feature enables you to access, display, and manipulate the project data in a variety of formats to perform schedule, cost, and resource analyses. You can perform these operations by using options in the **Tracking** window.

Choose the **Tracking** button from the Directory bar; the **Tracking** window will be displayed, as shown in Figure 7-7. Depending on the type of layout selected, two or more panes are displayed in the **Tracking** window. The left pane displays the **Projects** window with the WBS details. The upper right pane displays all projects of the selected node in the **Display: Projects** area. The lower right pane displays the profile of the selected project. The upper right pane displays the **Options Bar** with various options that can be used to display different dialog boxes. You can hide and unhide the displayed panes by choosing the required options from the **View** tab in the menubar, as shown in Figure 7-8.

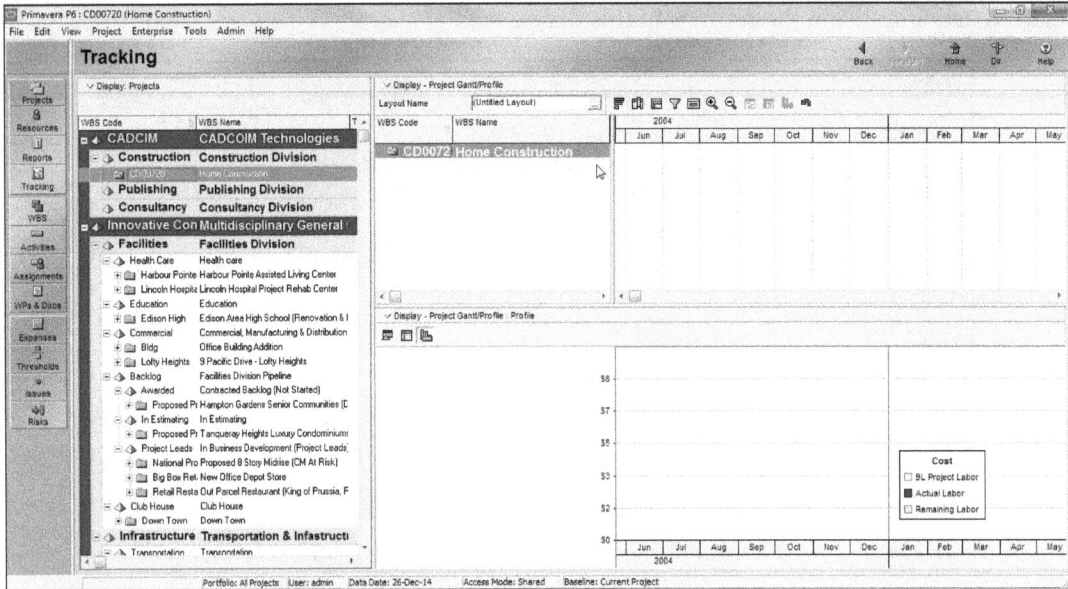

*Figure 7-7 The **Tracking** window*

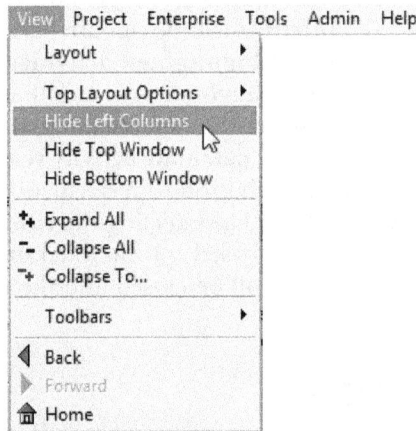

*Figure 7-8 Selecting the hiding options from the **View** tab in the menubar*

Working with Tracking Layouts

The display of the **Tracking** window depends on the layout selected. You can select the layout as per your requirement by choosing the Browse button next to the **Layout Name** edit box of the **Display- Project Gantt/Profile** pane. On doing so, the **Primavera P6** message box will be displayed prompting you to save the changes to this layout. If you want to save the changes to the layout then choose the **Yes** button; the **Save Layout As** dialog box will be displayed, as shown in Figure 7-9. In this dialog box, enter the name of the layout in the **Layout Name** edit box and then select the required option from the **Available to** drop-down list to define the user who can access the layout. If you select the **Current User** option from the **Available to** drop-down list only the admin administrator user can access the layout. Select the **All Users** option from the drop-down list to allow all the users to access the layout. Select the **Another User** option to allow the specified users to access the layout. Choose the **Save** button; the created layout is saved and the **Open Layout** dialog box will be displayed, as shown in Figure 7-10. If you choose the **No** button from the **Primavera P6** message box, the current layout will not be saved and the **Open Layout** dialog box will be displayed. In the **Open Layout** dialog box, you can choose the required option for the layout to be displayed. Then, choose the **Open** button; the selected layout type will be displayed, as shown in Figure 7-11.

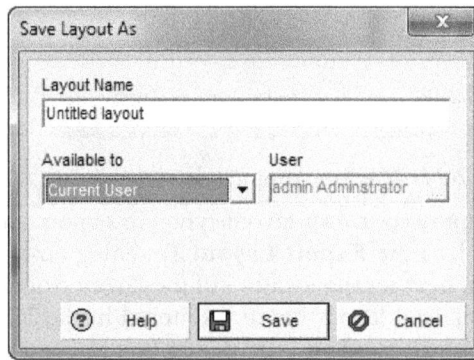

*Figure 7-9 The **Save Layout As** dialog box*

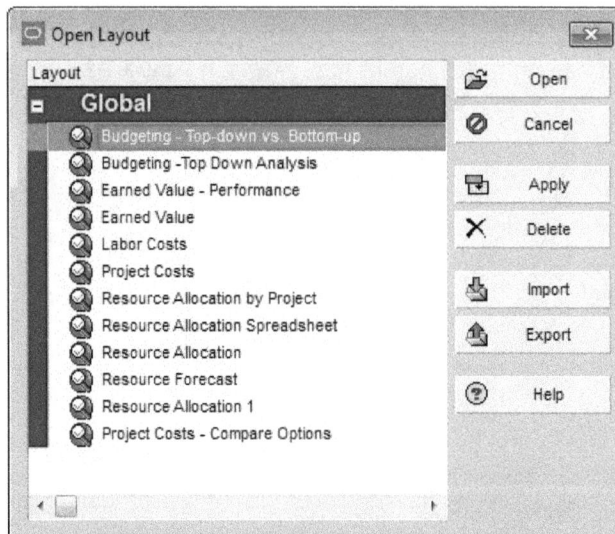

*Figure 7-10 The **Open Layout** dialog box*

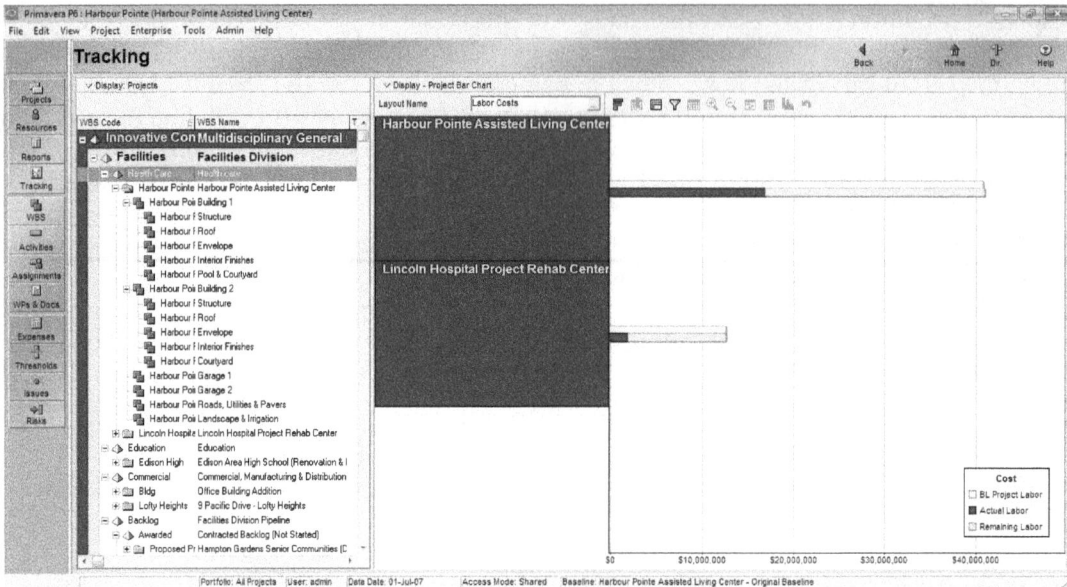

Figure 7-11 *The **Tracking** window with specified layout type*

You can also import and export your own layout type. To export a layout, choose the **Export** button from the Command bar; the **Export Layout To** dialog box will be displayed, as shown in Figure 7-12. In this dialog box, set the location of the layout from the **Save in** drop-down list to export the layout. The selected layout name is entered in the **File Name** edit box. Choose the **Save** button; the layout is exported and the dialog box will be closed.

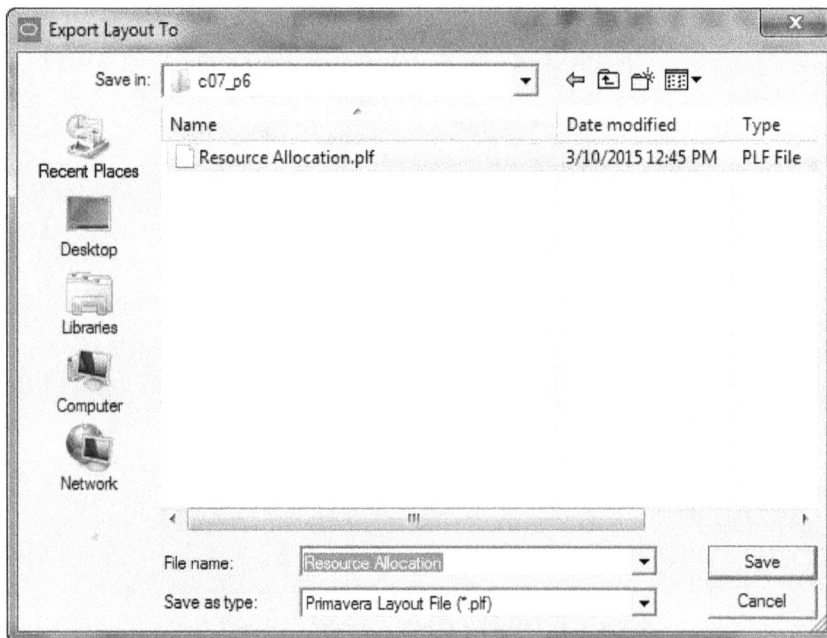

Figure 7-12 *The **Export Layout To** dialog box*

Similarly, you can import a layout. To do so, choose the **Import** button from the Command bar; the **Import Layout** dialog box will be displayed, as shown in Figure 7-13. In this dialog box, using the **Look in** drop-down list, browse to the location from where the file is to be imported. Enter the name of file to be imported in the **File name** edit box. Next, choose the **Open** button; the **Import Layout As** dialog box will be displayed. In this dialog box, the imported layout name is entered in the **Layout Name** edit box. Choose the required option from the **Available to** drop-down list to allow the user to access the imported layout. Now, choose the **Save** button; the imported layout is saved in the layout types. If you want delete a layout, choose the **Delete** button from the Command bar.

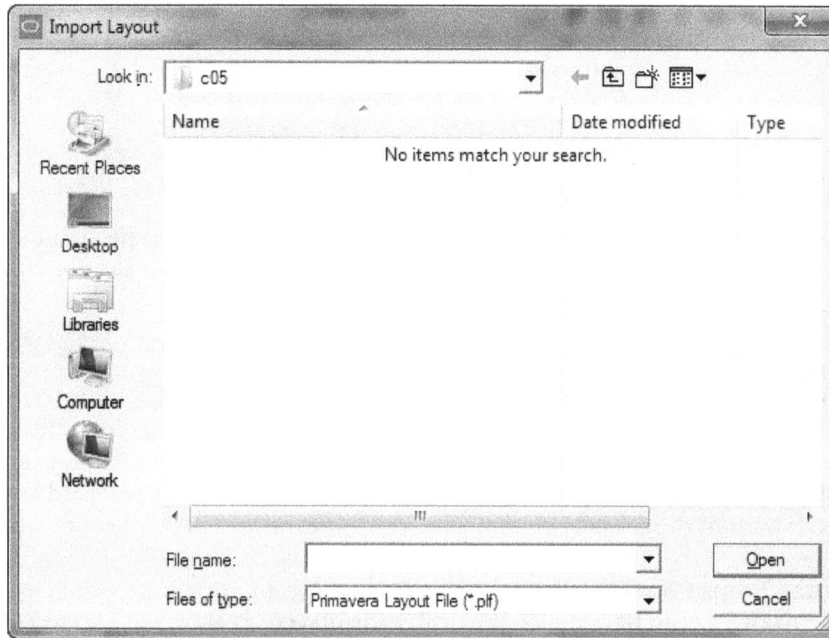

Figure 7-13 The Import Layout dialog box

Creating Tracking Layout

In Primavera, you can create a new layout. To do so, choose **Layout > New** option from the **Display - Project Gantt/Profile** drop-down; the **New Layout** dialog box will be displayed, as shown in Figure 7-14. In this dialog box, enter the name of the layout in the **Layout Name** edit box. You can select the required option from the **Available to** drop-down list to allow the selected users to access the layout. In the **Select Display Type** area, select the required radio button for the type of layout you want to display.

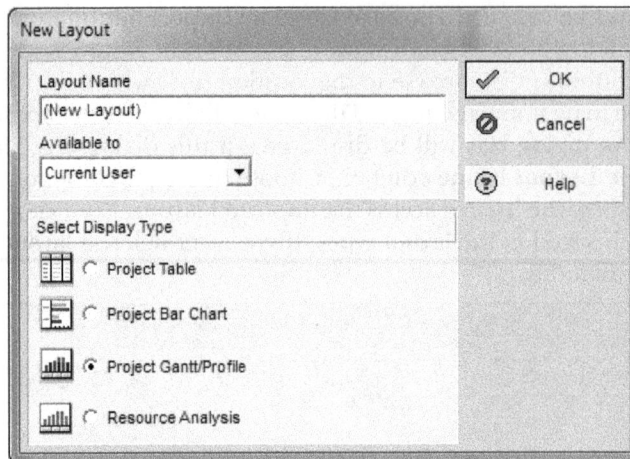

*Figure 7-14 The **New Layout** dialog box*

There are four types of radio buttons available in the **Select Display Type** area for displaying different types of layouts: **Project Table**, **Project Bar Chart**, **Project Gantt/Profile**, and **Resource Analysis**.

Select the **Project Table** radio button and then choose the **OK** button; the project data will be displayed in the table format layout and the **Columns** dialog box will be displayed. You can customize the columns of the project which are displayed in the right pane of this dialog box. In the **Columns** dialog box, select the required options that you want to display in the project table layout from the **Available Options** area and then move them to the **Selected Options** area. Choose the **OK** button; the selected columns will be displayed and the dialog box will be closed.

If you select the **Project Bar Chart** radio button, the project data will be displayed in the bar chart format layout and the **Bars** dialog box will be displayed, as shown in Figure 7-15. You can use the **Bars** dialog box to change the bar chart settings for the current layout. In this dialog box, select the WBS category from the **Display** drop-down list that you want to display in the project bar chart. Depending upon the selected WBS field, the options in the **Field** drop-down list will be modified in the **Show bars** area. You can select any option from the respective fields as per your requirement. You can change the color of the fields by choosing the button corresponding to the **Field** drop-down list. You can select the check boxes under the **Show Stacked** column corresponding to the **Field** drop-down list. On selecting these check boxes, the bars will stack with each other. Choose the **OK** button; the **Bars** dialog box will be closed.

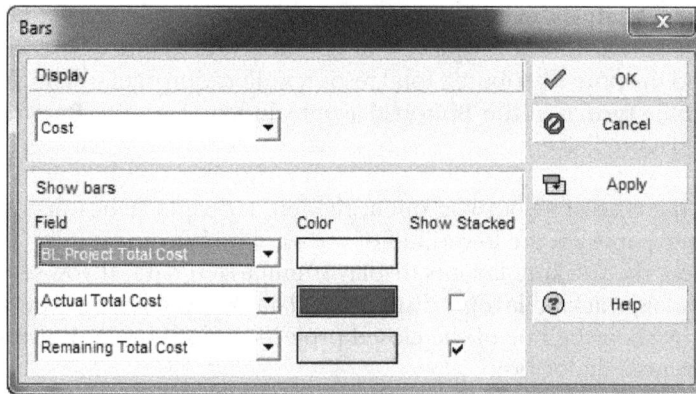

*Figure 7-15 The **Bars** dialog box*

If you select the **Project Gantt/Profile** radio button, the project information will be displayed in columns and Gantt chart format layout in the Top Layout window. And, the time-distributed project data will be displayed in either the spreadsheet or profile format in the Bottom Layout window. Also, the **Bars** dialog box will be displayed, as shown in Figure 7-16. The **Bars** dialog box helps you to specify the style and labels for the bars in a Gantt Chart. In this dialog box, select the check boxes under the **Display** column corresponding to the bars that you want to display in the bar chart. You can change the bar name according to your need by double-clicking in the corresponding **Name** column. Note that the bar name will not be displayed in the chart.

You can set the timescale of the bar by double-clicking in the **Timescale** column. The filter applied to the data item represented by the bar type can be edited by double-clicking in the **Filter** column. When you click on the **Filter** column, the **Filters** dialog box will be displayed. In this dialog box, you can change the filter type by selecting the required radio button.

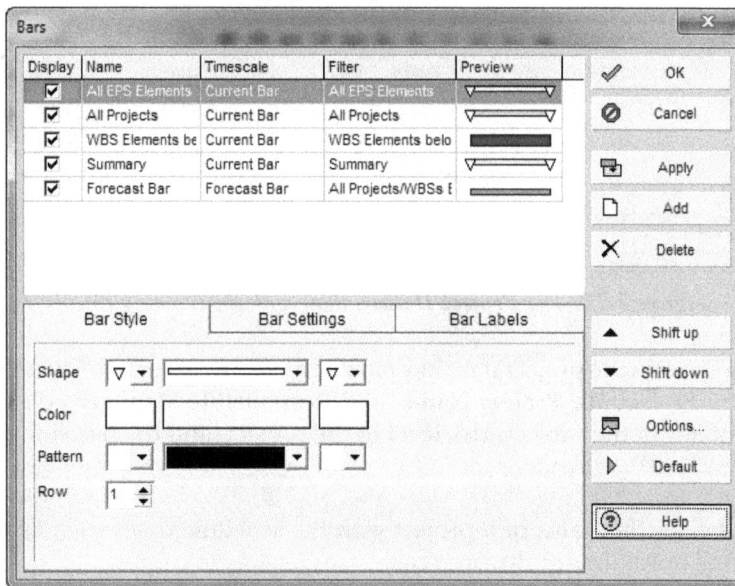

*Figure 7-16 The **Bars** dialog box*

Select the **Resource Analysis** radio button from the **New Layout** dialog box; the resource/ project usage information will be displayed in columns and Gantt Chart format in the Top Layout window, and the time-distributed total resource allocation data will be displayed in either spreadsheet or profile format in the Bottom Layout window. Also, the **Bars** dialog box will be displayed, refer to Figure 7-16.

Depending upon the type of layout you open, the left and right panes may split horizontally to display additional panes on the lower half of the window. When you select closed projects in the **Projects** window, the tracking layouts display summarized data. If you select open projects in the **Projects** window, tracking layouts display live data. You can change this setting to display summarized data by choosing one of the closed projects options in the **Resource Analysis** tab of the **User Preferences** dialog box.

Tracking Details Table

To view the additional information about a tracking project, right-click anywhere in the **Projects** window; a menu will be displayed. Choose the **Project Details** option from the menu; the **Project Details** table will be displayed, as shown in Figure 7-17. In this dialog box, four different tabs will be displayed with further information of the project. The detailed description of these tabs is given next.

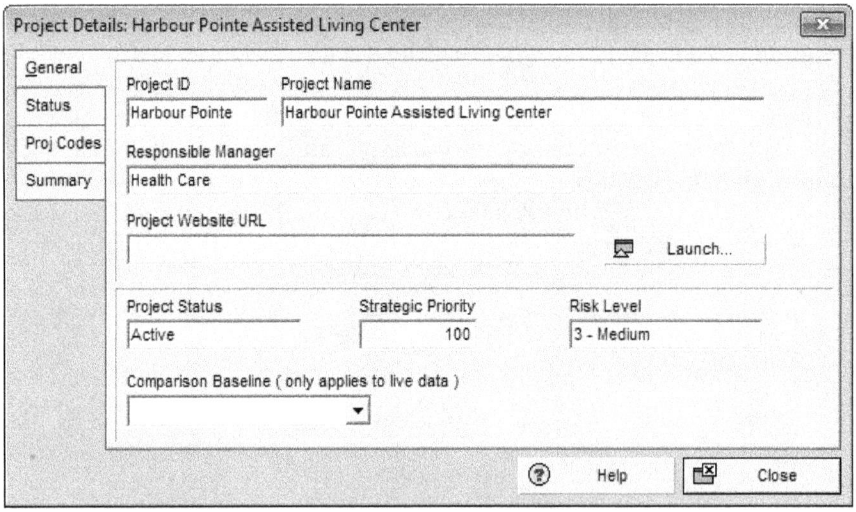

Figure 7-17 The Project Details table with the tracking details

The **General** tab displays the project data in a tabular form. In this tab, the details of the project are entered in the **Project ID**, **Project Name**, and **Responsible Manager** edit boxes. This tab also defines the status of the project, risk level of the project, and the baseline that is assigned to compare it.

The **Status** tab defines the status of a project with the schedule dates such as the actual start date and the actual finish date of the project.

The **Proj Codes** tab, as shown in Figure 7-18, specifies the code of the project with the code value and its description which is provided while creating a project.

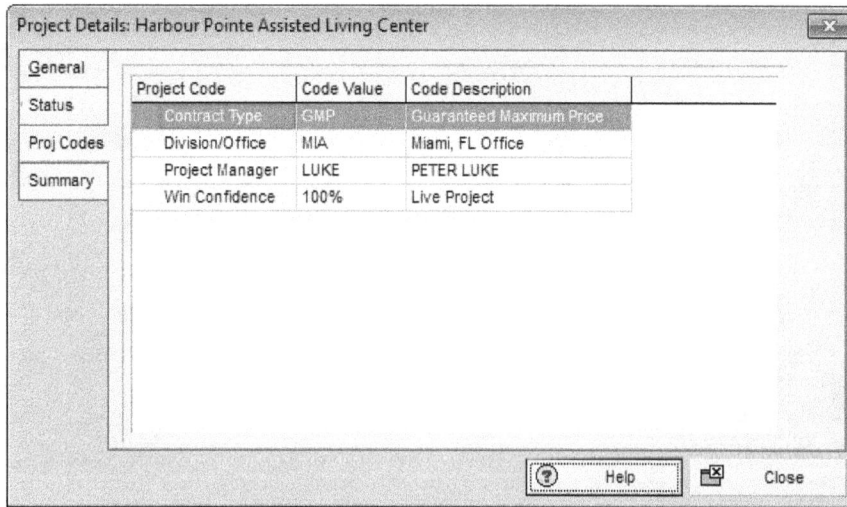

*Figure 7-18 The **Proj Codes** tab in the **Project Details** table*

The **Summary** tab displays the date on which the selected project was summarized last. This provides an idea to the users about how much current the summary data is. The **Summary** tab also displays information about the WBS level upto which the project was summarized. The **Last Summarized On** edit box displays the summary date on which the project was summarized. The WBS to be summarized is entered in the **Summarized to WBS Level** edit box.

GROUPING, SORTING, AND FILTERING DATA

Grouping data helps you to organize information into bands, based on a common attribute such as hierarchy, code value, or resource. You can choose to group data using the standard data groupings provided in the module, or you can create a customized grouping. Sorting enables you to determine the sequence of the data to be arranged in the layout.

You can group and sort the data of the layout according to your project. To do so, select the **Project** option from **Top Layout Options > Group and Sort By** in the **Display - Project Gantt/ Profile** drop-down; a list of project layout will be displayed. If you want to display EPS and WBS in the top layout window, select the **EPS/WBS** option from **Top Layout Options > Group and Sort By** in the **Display** drop-down; the EPS node with the WBS details will be displayed in the top layout window. If you want to display the project in the OBS format, select the **OBS** option from **Top Layout Options > Group and Sort By** in the **Display** drop-down; the OBS layout will be displayed. To display the layout according to the phase, select the **Phase** option from **Top Layout Options > Group and Sort By** in the **Display** drop-down; the layout with the phases will be displayed. Note that only those projects will be displayed to which project phase has been assigned. You can also customize the data groupings that can be applied to the layout. To customize the layout, select the **Customize** option from **Top Layout Options > Group and Sort By** in the **Display** drop-down; the **Group and Sort** dialog box will be displayed, as shown in Figure 7-19.

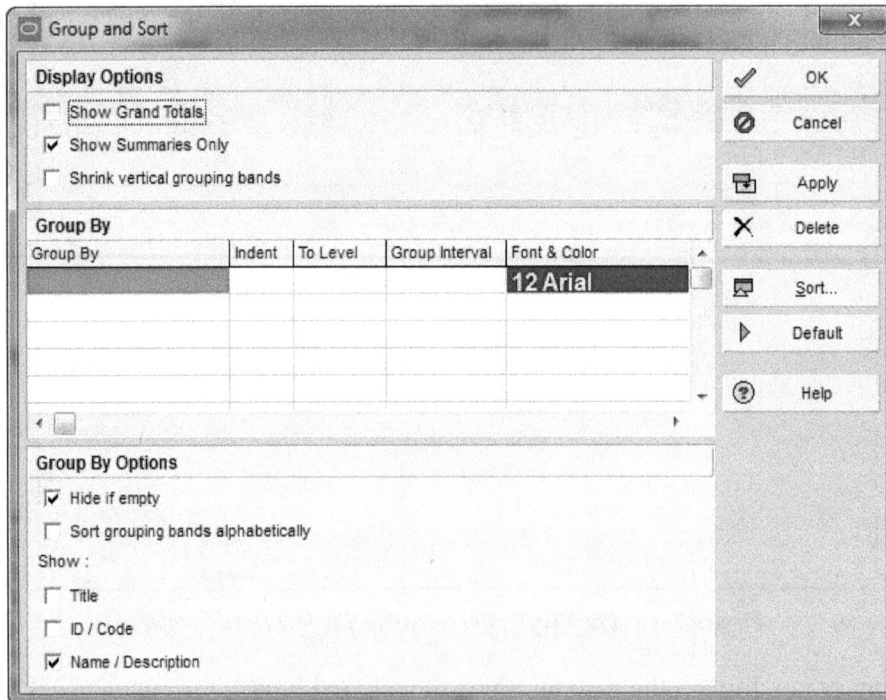

*Figure 7-19 The **Group and Sort** dialog box*

In this dialog box, you can select the **Show Grand Totals** check box in the **Display Options** area to display the layout with the grand total at the top of the layout, under which all the project data will be defined. Select the **Show Summarizes Only** check box to display the summarized information of project data in groups. Clear the check box to display the detailed description of project data in a group. Select the **Shrink vertical grouping bands** check box to shrink the height of vertical bands in the hierarchy.

In the **Group By** area, there are five columns. Double-click in the **Group By** column; a drop-down list will be displayed. Select the required option from the drop-down list to list the grouping items to be displayed. The **Indent** column gets automatically selected if the selected grouping item is hierarchical, as the hierarchy gets automatically indented. The **To Level** column helps to identify upto which level the hierarchy will be indented. The **Group Interval** column indicates the group interval between fields. For example, if in the **Group By** column, the **Remaining Duration** option is selected, then enter **5d** in the **Group Interval** column. On doing so, the data items in the project layout will be grouped according to the length of their remaining durations as 0.0d to 5.0d, 5.0d to 10.0d, and so on. You can customize the font and color of the text written in reference to the details of the project data by double-clicking in the **Font & Color** column. On doing so, the **Edit Font & Color** dialog box will be displayed, as shown in Figure 7-20. To change the font of the text, choose the **Font** button; the **Font** dialog box will be displayed, as shown in Figure 7-21. In

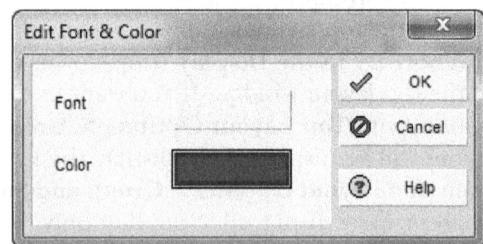

*Figure 7-20 The **Edit Font & Color** dialog box*

this dialog box, select the desired option under the **Font**, **Font Style**, and **Size** list. You can also apply effects such as **Strikeout** and **Underline** to the text by selecting the corresponding check boxes. You can change the color of the font by using the option in the **Color** drop-down list. Choose the **OK** button to close the **Font** dialog box. To change the color of the band of project layout, choose the **Color** button; the **Color** dialog box will be displayed. Choose the required color and then choose the **OK** button to close this dialog box.

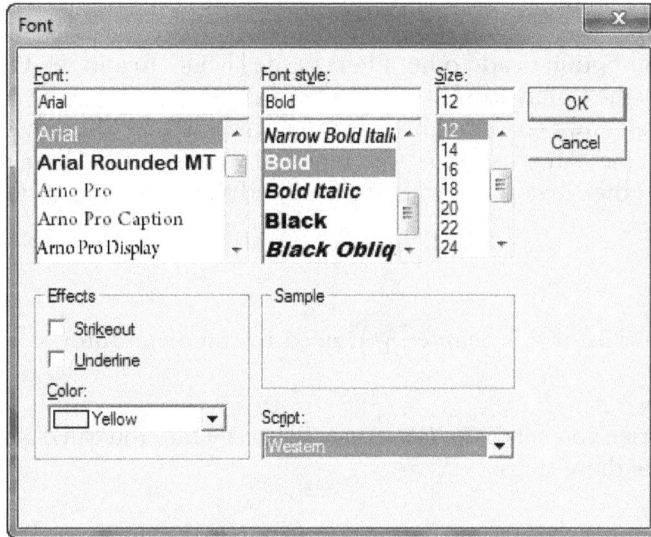

Figure 7-21 *The **Font** dialog box*

You can filter the data as per your requirement before displaying it in the layout. The filter is applied on the desired data to constrain it. To filter the layout, choose the **Filters** option from the cascading menu of the **Top Layout Options** in the **Display** drop-down; the **Filter** dialog box will be displayed, as shown in Figure 7-22.

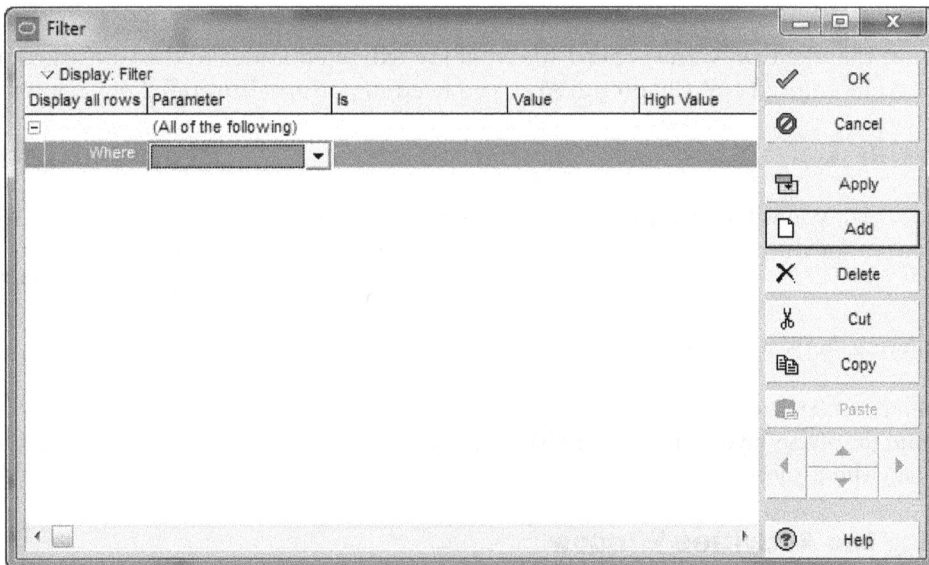

Figure 7-22 *The **Filter** dialog box*

In this dialog box, you can list each filter criteria item under the **Display all rows** column. The **Where** option is selected by default under this column. To list the parameters for each filter criteria item, select the required option from the **Parameter** drop-down list. You can assign the equal to, greater than or equals to, and other operations to each criteria item. To select the operator, double-click under the **Is** column, a drop-down list will be displayed. Select the required option from the displayed drop-down list. Enter the required value under the **Value** column depending upon the selected operator from the **Is** column.

You can use the **Filter** option to add other filters in the layout. To add the **Filter** option, choose the **Add** button from the Command bar; a new filter will be added with the **And** option selected under the **Display all rows** column. You can delete the filter by choosing the **Delete** button from the Command bar. After assigning the required values in the **Filter** dialog box, choose the **OK** button; the **Filter** dialog box will be closed and the layout will be filtered according to the provided settings.

TUTORIAL

To perform the tutorials of this chapter, you need to complete Tutorial 1 and Tutorial 2 of Chapter 2.

To perform the tutorial, you need *c06_Primavera_P6_tut.zip* file. You can download this file from *www.cadcim.com* using these steps.

1. To download the file, browse to *Textbooks > Civil/GIS> Primavera P6 > Exploring Oracle Primavera P6 v7.0*. Next, select the *c06_Primavera_P6_tut.zip* file from the **Tutorial Files** drop-down list. Choose the corresponding **Download** button to download the data file.

2. Now, save and extract the downloaded folder to the following location:

    C:/PM6

3. Import the *c06_CONS_Home_tut01* file from the extracted folder to the Construction EPS created in Tutorial 1 of Chapter 2. Also, you need to perform Tutorial 2 of Chapter 6.

Tutorial 1   Project Status

In this tutorial, you will create the project status for the **Home Construction** project. You will set the layout for the project and then will update the project with the new data date.
(Expected time: 45 min)

The following steps are required to complete this tutorial:

a. Open the **Activities** window.
b. Create the status layout in the **Activities** window.
c. Update the project status.

Opening the Activities Window

1. Open Primavera P6 and then open the **Projects** window by choosing the **Projects** option from the **Enterprise** menu in the menubar.

2. In this window, select the **Home Construction** project placed under the **Construction** EPS node and right-click; a shortcut menu is displayed.

3. Choose the **Open Project** option from this menu; the **Activities** window is displayed.

Creating the Status Layout

1. In the **Activities** window, choose the **Columns** button from the **Activities** toolbar; the **Columns** dialog box is displayed, as shown in Figure 7-23.

2. In this dialog box, select the **Activity Type** and **Original Duration** options from the **Selected Options** area and then choose the **Remove from list** button to remove the selected options.

3. Now, in the **Available Options** area of this dialog box, select the **Actual Start**, **Actual Finish**, **BL Project Start**, and **BL Project Finish** options under the **Dates** node and then move them to the **Selected Option** area using the **Add to list** button.

4. Similarly, in the **Available Options** area, select the **Activity Status** option under the **General** node and move it to the **Selected Options** area.

5. Now, select the **Remaining Duration** option under the **Durations** node and **Activity %** **Complete** option under the **Percent Completes** node in the **Available Options** area and move them to the **Selected Options** area using the **Add to list** button. Using the **Move up in list** and **Move down in list** buttons arrange these options, refer to Figure 7-23.

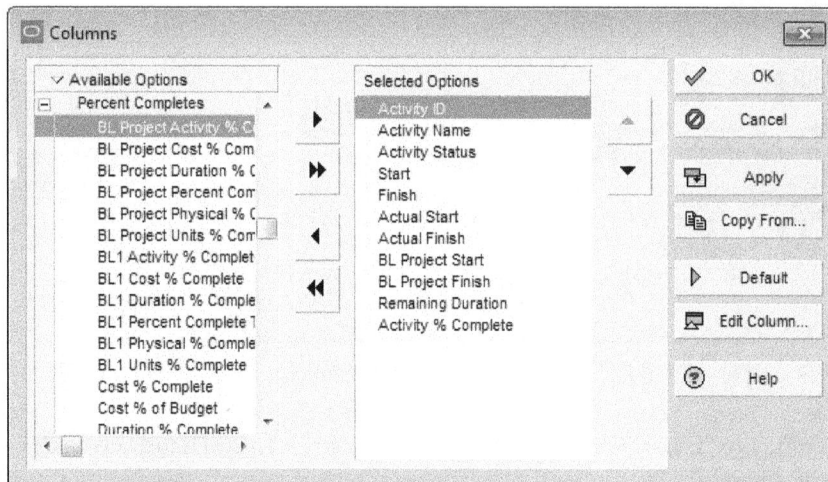

*Figure 7-23 The **Columns** dialog box*

6. Now, choose the **OK** button; the dialog box is closed and the selected columns are displayed in the **Activities** window.

7. Choose **Layout > Save As** option from the **View** menu of the menubar; the **Save Layout As** dialog box is displayed.

8. In this dialog box, enter **Status Layout** in the **Layout Name** edit box and then select the **All Users** option from the **Available to** drop-down list.

9. Choose the **Save** button; the layout with the assigned name is displayed.

Updating the Project Status

1. To update the status after week 1, you need to enter the following information in the **Activities** window under the columns shown in Table 7-1 below.

Table 7-1 *The details to be updated in a project*

Activity ID	Activity Name	Activity Status	Activity % Complete
A1000	Project Start	Completed	100%
A1030	Survey and Mark Out Site	Completed	100%
A1040	Site Demarcation	Completed	100%
A1050	Excavation	Completed	100%
A1040	Grade Site	In Progress	60%

2. Now, select the **Schedule** option from the **Tools** menu of the menubar; the **Schedule** dialog box is displayed, as shown in Figure 7-24.

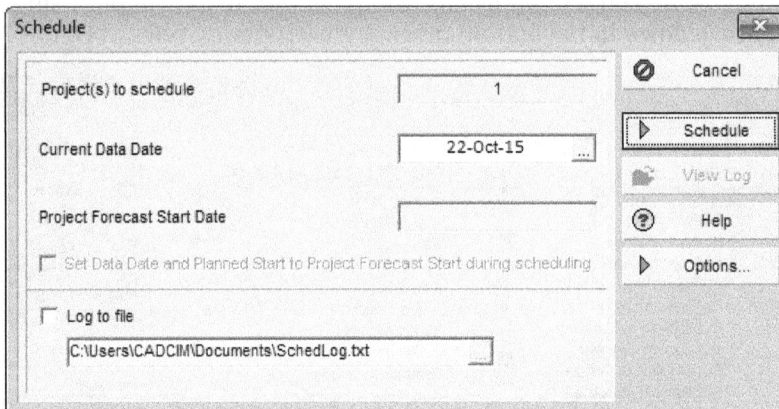

Figure 7-24 The **Schedule** dialog box with the scheduled date

3. In this dialog box, enter the date as **22-Oct-15** in the **Current Data Date** edit box and then choose the **Schedule** button, refer to Figure 7-24; the project is scheduled and updated. The updated project is displayed with the Gantt chart, as shown in Figure 7-25.

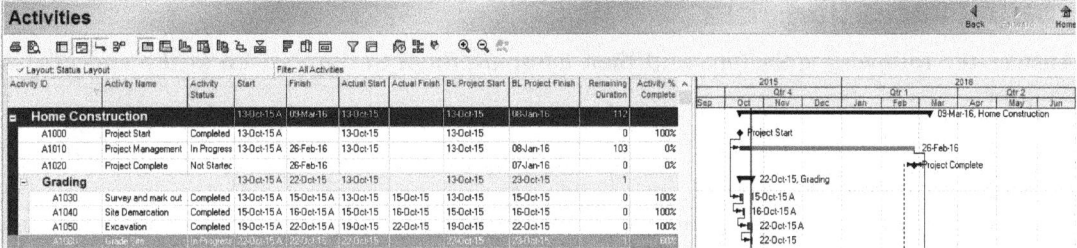

*Figure 7-25 The **Activities window** with the scheduled project in the Gantt chart*

4. Now, in the **Activities** window, choose the **Save As** option from the **Layout** drop-down; the **Save Layout As** dialog box is displayed.

5. In this dialog box, enter **Status Layout_week 2** in the **Layout Name** edit box and then choose the **Save** button; the **Status Layout_week2** is displayed in the **Layout** drop-down.

6. Now, update the project status data for week 2 with the information given in Table 7-2 below.

Table 7-2 The details to be updated in project for week 2

Activity ID	Activity Name	Activity Status	Activity % Complete
A1060	Grade Site	Completed	100%
A1070	Laying of Foundation	Completed	100%
A1080	Slab Plumbing	In Progress	50%

7. Schedule the project by choosing the **Schedule** option from the **Tools** menu; the **Schedule** dialog box is displayed. In this dialog box, enter the date **24-Oct-15** in the **Current Data Date** edit box.

8. Choose the **Schedule** button; the project is updated and is displayed with the Gantt chart, as shown in Figure 7-26.

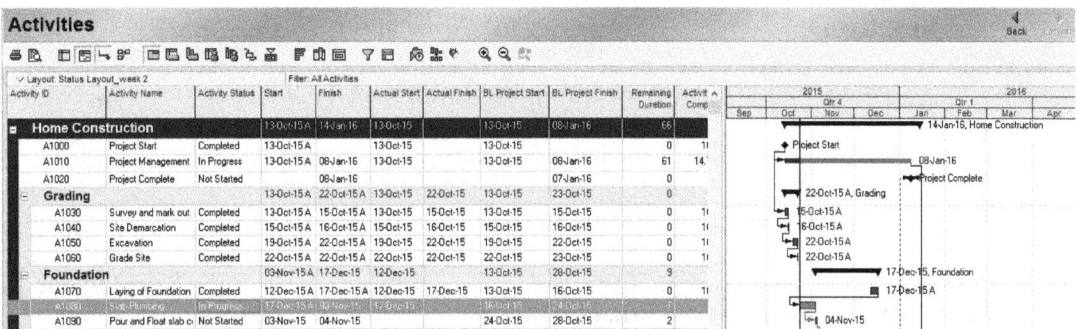

*Figure 7-26 The **Activities** window with the scheduled project in the Gantt chart*

Self-Evaluation Test

Answer the following questions and then compare them to those given at the end of this chapter:

1.   In the _____ dialog box, you can update the duration and actual values of the activities.

2.   The ____, ____, _____, and ____ are the project status types that can be assigned according to the progress of project.

3.   You can set the name of the layout in the _____ edit box.

4.   The created layout can be saved in the _____ dialog box.

5.   The settings of layout can decide the display of the window. (T/F)

6.   The options in the **Project Status** drop-down list allow you to set the status. (T/F)

7.   You can display only one pane in the **Tracking** window. (T/F)

8.   The **Bars** dialog box can be used to set the type and color of bar. (T/F)

Review Questions

Answer the following questions:

1.   Which of the following dialog boxes allows you to update the progress?

   (a) **Schedule**   (b) **Update Progress**
   (c) **Apply Actuals**   (d) **Level Resources**

2.   Which of the following dialog boxes allows you to open the created layout?

   (a) **Open Layout**   (b) **Save Layout As**
   (c) **New Layout**   (d) None of these

3.   The _____ radio button allows you to display project data in tabular format.

4.   You can set the shape and size of the bar using the _____ dialog box.

5.   The _____ option allows you to group and sort data in the layout.

6.   You can choose the desired option from the _____ drop-down list of the **Save Layout As** dialog box to allow users to access that layout.

7. The layout can be exported using the **Export** dialog box for further use. (T/F)

8. The **Open Layout** dialog box allows you to select the layout type to be displayed in the window. (T/F)

9. The timescale of the bar can be set from the **Timescale** column of the **Bars** dialog box. (T/F)

10. The project status can be updated using the **Schedule** dialog box. (T/F)

EXERCISE

Exercise 1

In this exercise, you will set the layout for the **Hospital Building** project and update the project with different data dates.   **(Expected time: 45 min)**

Hints:
1. Update the status for week 1.
2. Similarly, update the status of other weeks.

Answers to Self-Evaluation Test

1. Update Progress, 2. Active, Inactive, Planned, What if, 3. Layout Name, 4. Save Layout As, 5. T, 6. T, 7. F, 8. T

Chapter 8

Printing Layouts and Reports

Learning Objectives

After completing this chapter, you will be able to:

- *Understand reports*
- *Understand type of reports*
- *Understand the printing concept*

INTRODUCTION

In the previous chapter, you learned to set layout for the **Activities** and **Projects** windows and also to update the project status according to the progress of the project. Also, you learned the tracking of projects by creating and maintaining baselines.

In this chapter, you will learn to print layouts and reports for the distribution of project data. This chapter focuses on the page settings (such as margins, page orientation, header, and footer settings, and so on), preview layout, and printing. Moreover, you will learn to open standard reports, create new reports, and modify existing reports.

REPORTS

Reporting is the process of monitoring a project. It helps in communicating the project progress with the team members and the executive management. Generating reports is a useful way for keeping track of important information.

Reports help business managers to take better decisions. It can be presented in two forms: tabular and graphical. The report types will be explained later in this chapter. Firstly, you will learn to create and modify reports and report groups.

Creating Reports

You can create and modify reports in **Report Wizard**. To open **Report Wizard**, choose the **Reports** button from the Directory bar; the **Reports** window will be displayed. In this window, you can create a report by using the **Add** button from the Command bar. When you choose this button, **Report Wizard** with the **Create or Modify Report** page will be displayed, as shown in Figure 8-1.

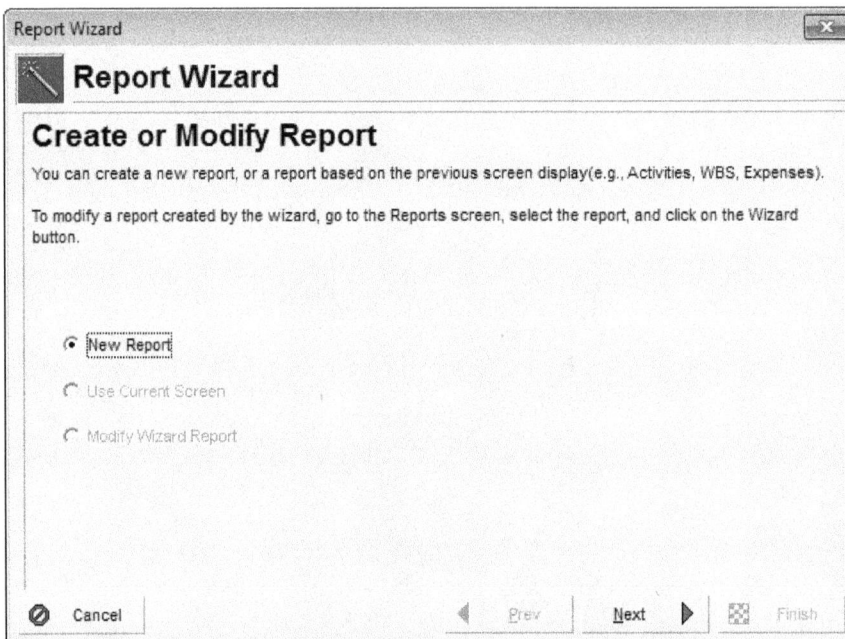

*Figure 8-1 The **Report Wizard** with the **Create or Modify Report** page*

In this page, the **New Report** radio button is selected by default which allows you to create a new report. Now, choose the **Next** button; the **Select Subject Area** page will be displayed. In this page, select the desired subject areas for the report and then choose the **Next** button; the **Select Additional Subject Areas** page will be displayed, as shown in Figure 8-2. In this page, select any desired additional option from the **Available Subject Areas** area and choose the arrow button to move them to the **Selected Subject Areas** to be added to the report. Now, choose the **Next** button; the **Configure Selected Subject Areas** page will be displayed, as shown in Figure 8-3. Note that if you select the **Time Distributed Data** check box in the **Select Subject Area** page, the subject areas related to time will be displayed. You can select the desired option from the area and then choose the **Next** button to display the **Configure Selected Subject Areas** page.

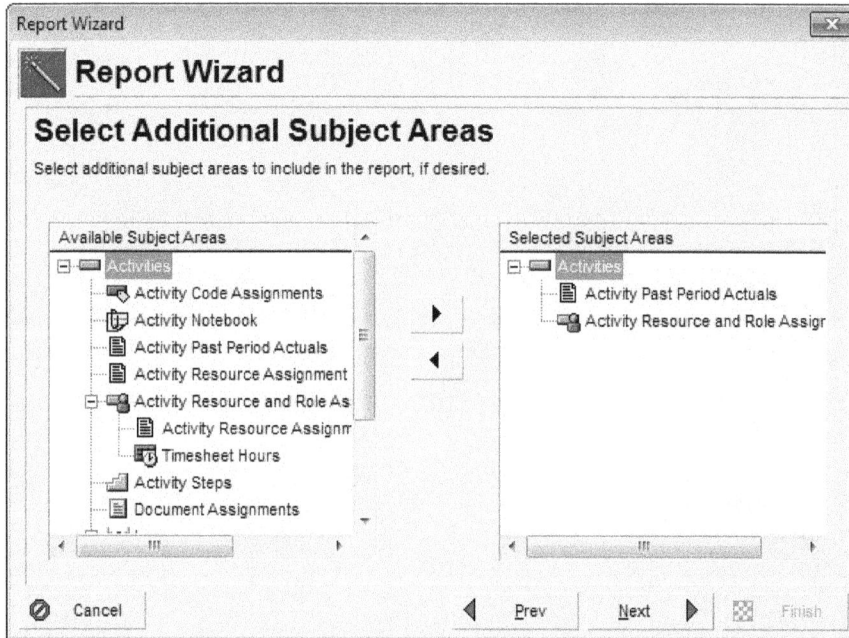

Figure 8-2 Report Wizard with the Select Additional Subject Areas page

In the **Configure Selected Subject Areas** page, you can configure the columns for the selected subject area which will be displayed in the report. To configure the columns, choose the **Columns** button from the **Activities** area; the **Columns** dialog box will be displayed. In this dialog box, select the options to be displayed in the report from the **Available Options** area and move them to the **Selected Options** area.

Similarly, you can choose the **Group & Sort** button for grouping and sorting the selected subject area. On choosing this button from the **Configure Selected Subject Areas** page, the **Group & Sort** dialog box will be displayed, as shown in Figure 8-4. In this dialog box, select the option from the **Group By** column by which the grouping is to be done. Select the level entry upto which level the report is to be grouped from the **To Level** column. Also, you can define the font and color for the report from the **Font & Color** column. In the **Group By Options** area, you can select the required option for showing the total of the required report from the **Show Totals** drop-down list. Now, choose the **Apply** and then the **OK** button; the dialog box will be closed and the changes will be saved.

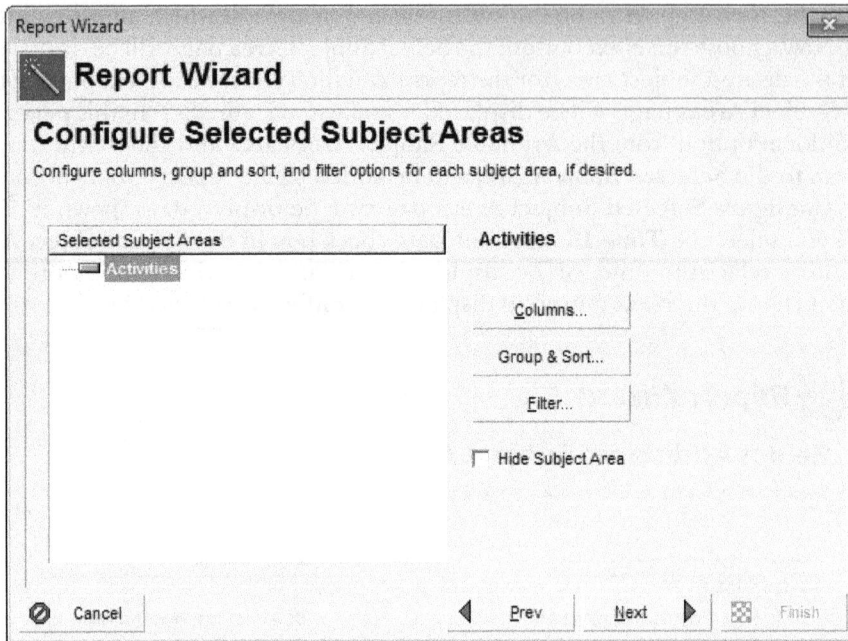

*Figure 8-3 The **Report Wizard** with the **Configure Selected Subject Areas** page*

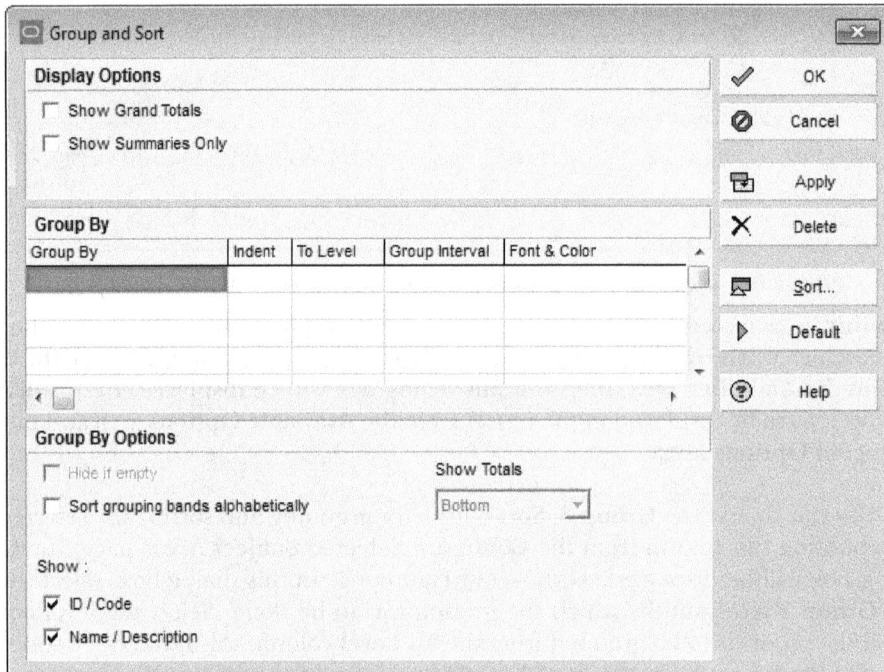

*Figure 8-4 The **Group and Sort** dialog box*

If you want to filter reports then you can choose the **Filter** button from the **Configure Selected Subject Areas** page. On doing so, the **Filter** dialog box will be displayed, as shown in Figure 8-5.

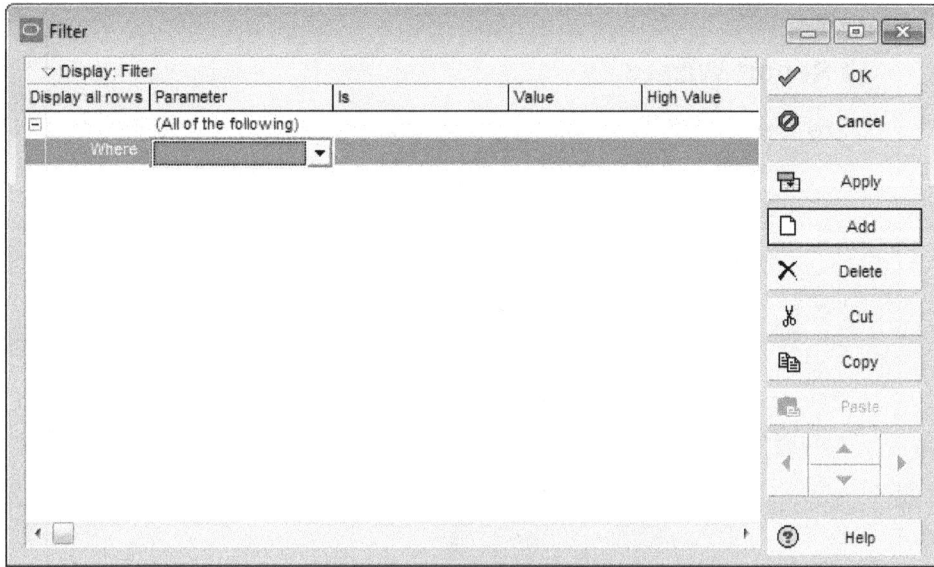

Figure 8-5 *The **Filter** dialog box*

In this dialog box, select the required option from the **Parameter** column according to which the filtering is needed to be done. In the **Is** column, specify the constraints such as equals, is greater than, is greater than or equals to, and so on and then assign a value in the **Value** column. Enter the highest value in the **High Value** column if you want strict filtering. Then, choose the **OK** button; the dialog box will be closed and the filter in the report will be applied. Now, choose the **Next** button from the **Configure Selected Subject Areas** page; the **Report Title** page will be displayed. Note that if the **Time Distributed Data** check box is selected in the **Select Subject Area** page, then on choosing the **Next** button from the **Configure Selected Subject Areas** page; the **Date Options** page will be displayed, as shown in Figure 8-6.

In the **Date Options** page, the **Timescale** and the **Time Interval Fields** buttons are displayed. These two buttons are displayed for viewing the data over time. Select the **Timescale** button; the **Timescale** dialog box will be displayed, as shown in Figure 8-7.

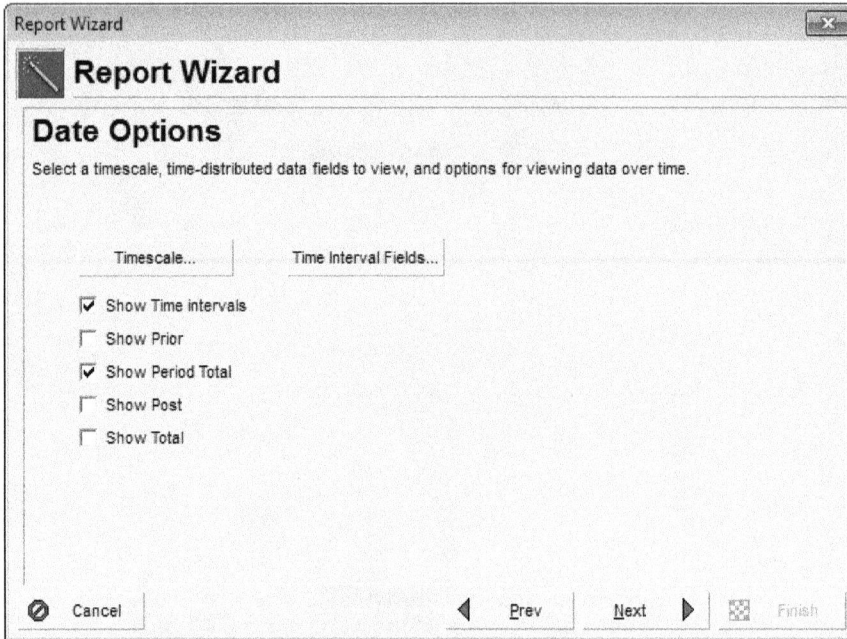

Figure 8-6 *The* ***Date Options*** *page of the* ***Report Wizard***

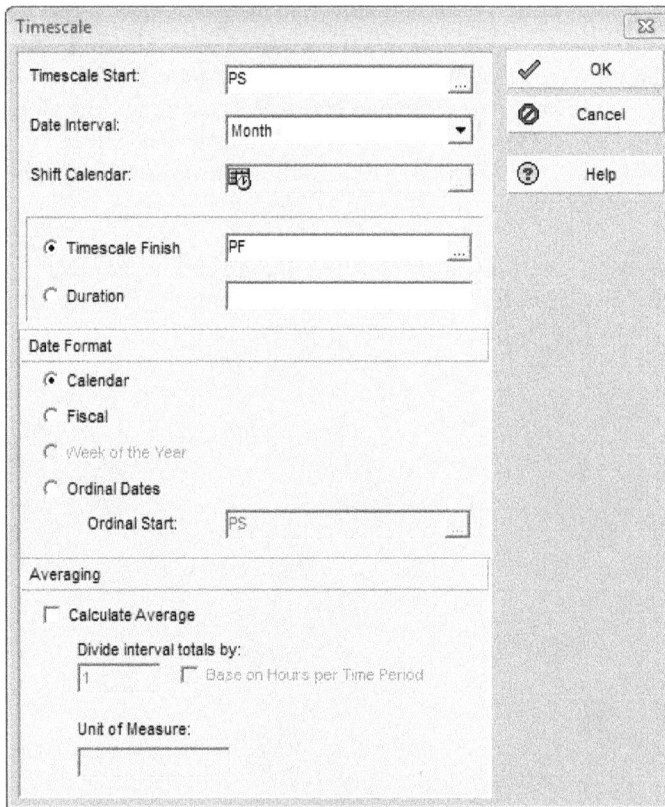

Figure 8-7 *The* ***Timescale*** *dialog box*

In this dialog box, you will provide the necessary details of the time for the report generation. In the **Timescale Start** field, choose the Browse button; a menu will be displayed. In this menu, select the required date for the report to be displayed on. The **PS** option is selected by default. In the **Date Interval** drop-down list, select an option according to which the interval of the report will be displayed. In the **Timescale Finish** field, the **PF** option is selected by default. You can change the default value by choosing the Browse button from the field and then by choosing the desired option from the menu displayed. Select the **Duration** radio button to enable the corresponding edit box. In this edit box, you can manually enter the duration of the report interval to be generated. The **Date Format** area displays the format of dates that needs to be displayed in the report. Select the required radio button for date format that you wish to display in the report. In the **Averaging** area, you can select the **Calculate Average** check box to enable the options in the area. This allows to divide the timescale interval totals. In this area, you can enter the timescale total intervals in the **Divide interval totals by** edit box. Also, you can select the **Base on Hours per Time Period** check box to divide the timescale intervals by automatic intervals. In the **Unit of Measure** edit box, you are required to enter the unit for the timescale intervals. Now, choose the **OK** button from the **Timescale** dialog box; the dialog box will be closed and the **Date Options** page will be displayed again.

In the **Date Options** page, choose the **Time Interval Fields** button; the **Fields** dialog box will be displayed, as shown in Figure 8-8.

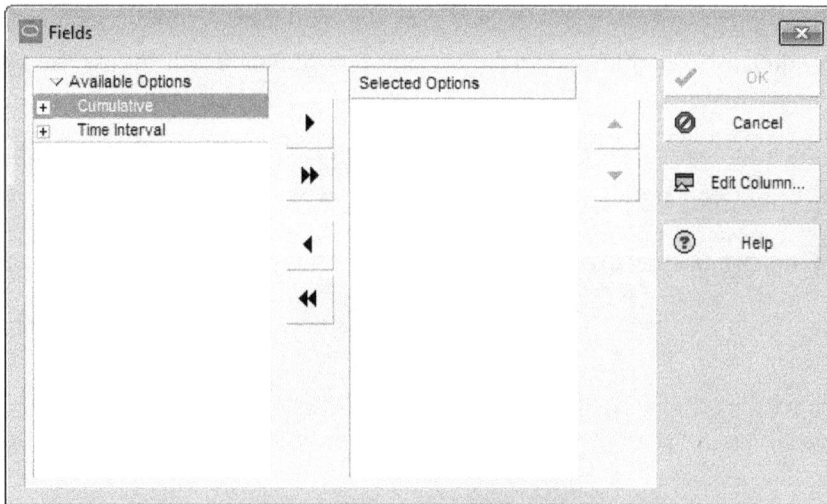

Figure 8-8 The Fields dialog box

Using this dialog box, you can include the field that is to be added in the spreadsheet layout. Select the required option from the **Available Options** area and shift them to the **Selected Options** area. Next, choose the **OK** button; the dialog box will be closed. Choose the **Next** button from the **Date Options** page; the **Report Title** page will be displayed, as shown in Figure 8-9.

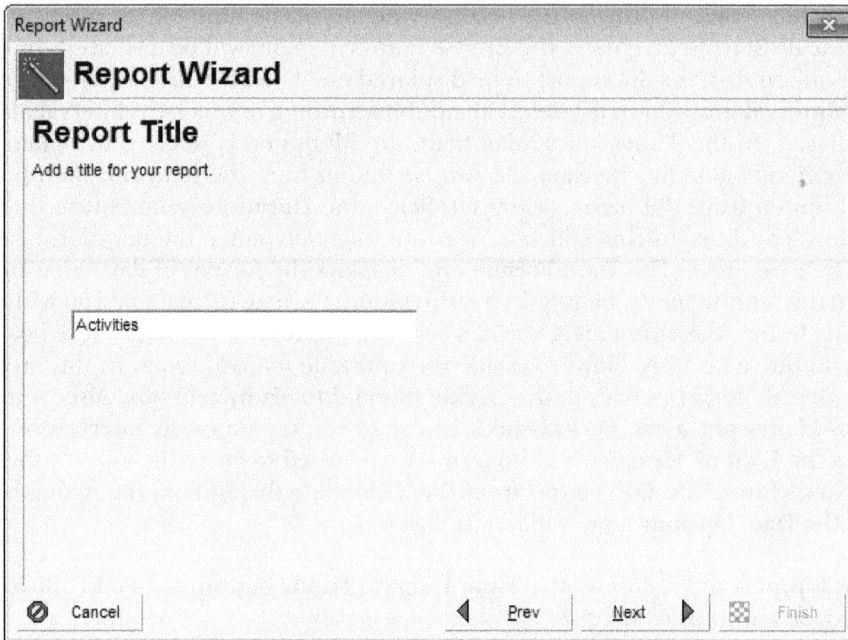

*Figure 8-9 The **Report Title** page of **Report Wizard***

In this page, enter the title of the report in the edit box provided and choose the **Next** button; the **Report Generated** page will be displayed, as shown in Figure 8-10.

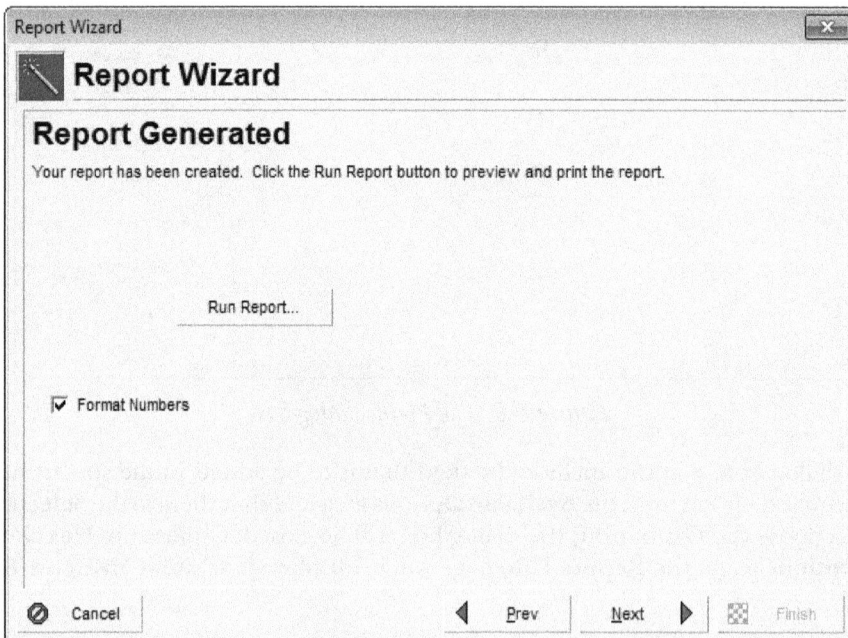

*Figure 8-10 The **Report Generated** page of **Report Wizard***

This page indicates that the report has been created. You can choose the **Run Report** button to run the report. Now, choose the **Next** button; the **Congratulations** page will be displayed, as shown in Figure 8-11.

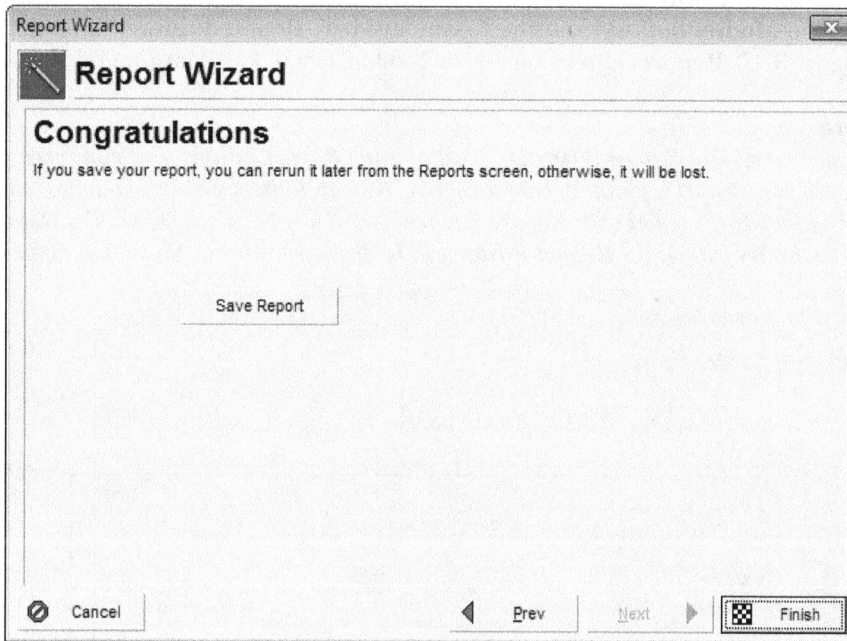

*Figure 8-11 The **Congratulations** page of **Report Wizard***

In this page, choose the **Save Report** button and then choose the **OK** button; the report will be saved in the **Reports** window. Next, choose the **Finish** button; the report will be produced with all the defined settings.

You can create new reports or can modify existing reports using **Reports Wizard** or **Report Editor**. **Report Wizard** allows you to create a wide variety of reports using a wizard style interface. You can also modify the reports created using **Reports Wizard** which is discussed in detail next.

Modifying Reports

You can modify a report by using the **Report Wizard** or **Report Editor**. To modify a report select the required report file from the **Reports** window and then choose the **Wizard** button from the Command bar; the **Create or Modify Report** page of **Report Wizard** will be displayed. In this page, the **Modify Wizard Report** radio button will be selected by default. Now, in **Report Wizard** make modification and then perform other processes as done earlier in creating reports.

Creating Report Using Report Editor

Report Editor is a report writer which allows you to group, sort and, roll up projects information. It includes HTML links in the reports. You can use **Report Editor** to customize the reports created in **Reports Wizard**. Therefore, if the wizard report is modified in **Report Editor** then on reopening the report in the wizard, all modifications created in **Report Editor** will be lost.

You can create a report using **Report Editor**. **Report Editor** enables you to create customized report and also to specify other required settings. **Report Editor** allows to create, edit, and organize the report component. The report component can be a data source, a row, or a cell. Now, to create a report with the editor, choose any report type from the **Reports** window and then choose the **Modify** button from the Command bar; **Report Editor** will be displayed, as shown in Figure 8-12. **Report Editor** consists of Toolbar, Scale, Left Margin and Report Canvas.

Note

*If reports created in **Report Wizard** is modified using **Report Editor**; the **Primavera P6** message box will be displayed informing that the changes done in **Report Editor** will be lost whenever you will again edit the report with **Report Wizard**. If you want to make changes in **Report Editor**, choose the **Yes** button; the **Report Editor** will be displayed. Choose **No** to discard the changes.*

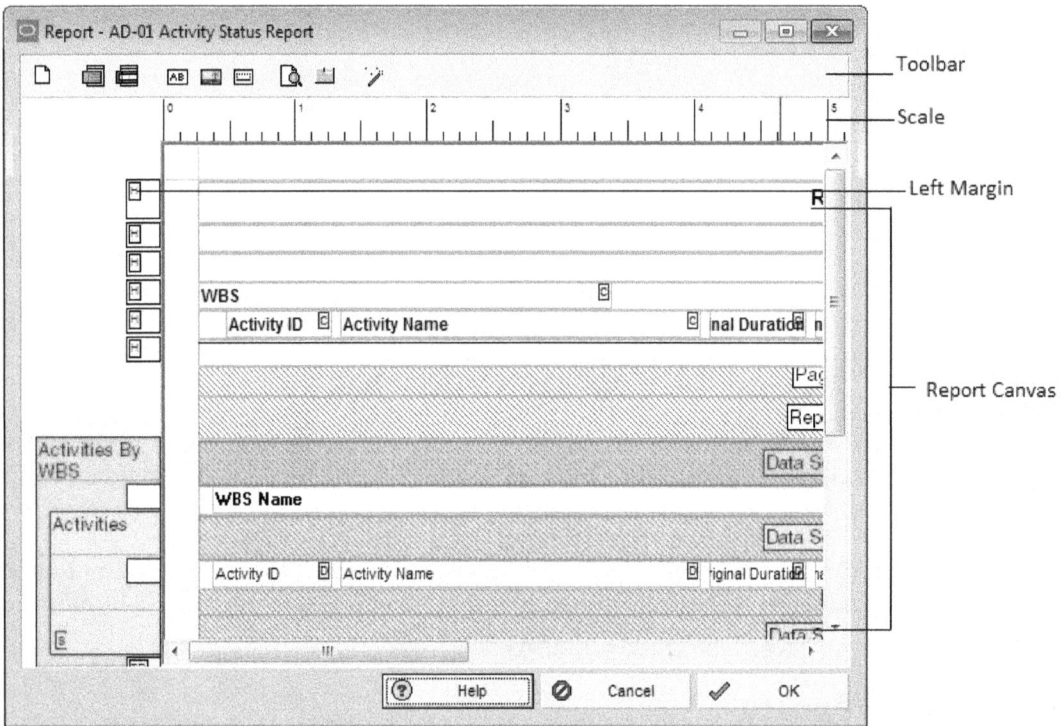

*Figure 8-12 The **Report Editor***

The Toolbar displays the tools for performing functions within **Report Editor**. In this Toolbar, there are tools to create new report headings, to add data source, to add rows to the report, and also to add text in the provided areas. You can also preview the report by choosing the **Preview** button from the Toolbar. The **Properties** button allows you to define the properties of the selected report component. To edit the properties of a component, double-click on it; the **Properties** dialog box will be displayed, as shown in Figure 8-13. You can also modify the current report settings using **Report Wizard**.

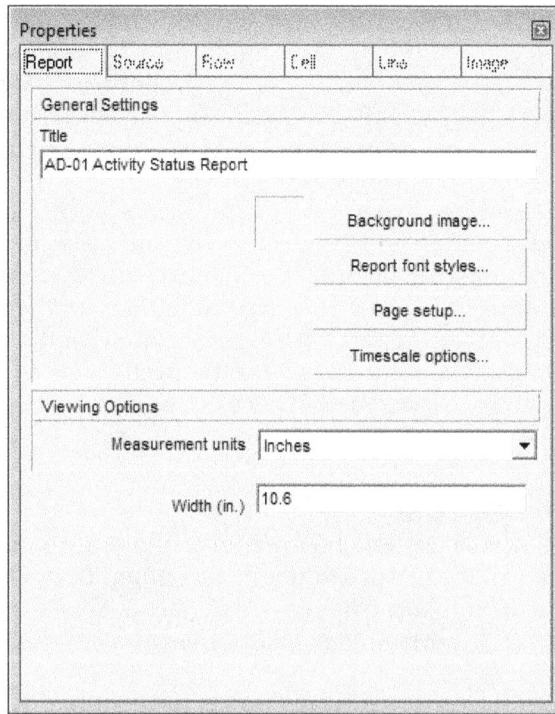

*Figure 8-13 The **Properties** dialog box*

The Scale defines the horizontal position of the report component. The blue area on the scale signifies the width of the report component.

Note

*On selecting WBS, Activity ID, and so on from **Report Editor**, the blue area will be highlighted on the scale that defines the width of an individual report component.*

The Left Margin displays the rows for each data source. **H** written on the block indicates that it is a part of the **Header** area and **F** written in front of the row indicates the **Footer** area. **C** written in front of the cell indicates that this is a **Custom Text** cell type and the text in this cell is specified by the user. **D** represents the **Field Data** cell type that compiles information from the location specified. **T** represents the **Field Title** cell type which represents that the name of a field is specified by the user. **V** represents the **Variable** cell type which is for that cell which reports information of the overall report.

The Report Canvas allows you to view the position of the component in overall report. It also provides the color codes to the components. In **Report Editor**, the text cells are coded according to the type of report data such as C, D, T, and V which has been discussed earlier.

Report Editor allows you to add a new report. To do so, choose the **New Report** tool from the Toolbar. On doing so, the **Primavera P6** dialog box will be displayed prompting you to clear the current report and to start over with a blank one. If you want to start over with a new report choose the **Yes** button and then create a new report as discussed earlier. You can select the **Add data source** tool to add the additional data source to the report. If you want to add data row to the report then choose the **Add row** tool from the Toolbar. While modifying or creating a report in **Report Editor**, you can add text cell by choosing the **Add Text cell** tool from the Toolbar to add any text under it. When you choose this tool, the **Properties** dialog box with the **Cell** tab chosen will be displayed. In the **Cell** tab, enter necessary details for the text cell that is to be added. You can add the image cell by using the **Add image cell** tool from the Toolbar. When you choose this tool, the **Properties** dialog box with the **Image** tab chosen will be displayed. You can add any image and can provide general settings for the image in the **Properties** dialog box. The image to be inserted should be in *.bmp file format. To insert the horizontal line cell in the report, choose the **Add line cell** tool from the Toolbar; the **Properties** dialog box with the **Line** tab selected will be displayed. Specify the general settings for the line cell. After creating and modifying the reports with the **Report Editor**, you can preview the reports by choosing the **Preview** tool from the Toolbar. You can choose the **Properties** tool to display the **Properties** dialog box in which you can select the required tab to make changes in it. Next, choose the **OK** button; the dialog box will be closed.

Creating Report Groups

Report group is a way to represent the hierarchy of reports to organize global or project report. Each report in the group belongs to one report group of its type. The report groups are maintained to place similar report types under one head. Some report groups are already created and displayed in the **Reports** window while others can be created.

To create the report group, select the **Report Groups** option from the **Reports** flyout of the **Tools** menu; the **Report Groups** dialog box will be displayed, as shown in Figure 8-14.

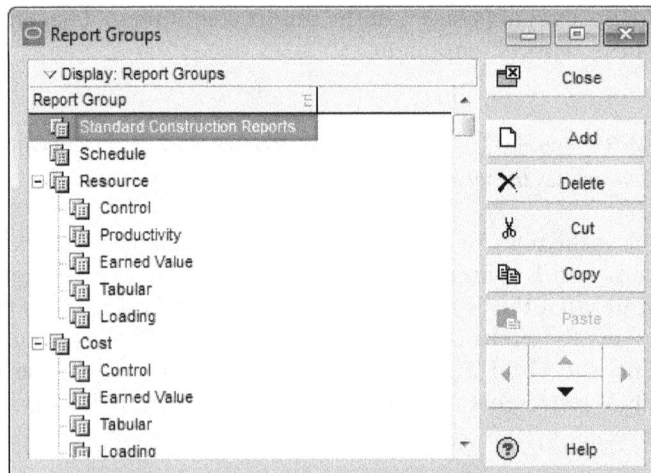

*Figure 8-14 The **Report Groups** dialog box*

In this dialog box, you can see the existing report groups which are displayed in the **Reports** window. To create a new report group, choose the **Add** button from the Command bar; a new report group with the name **New Report Group** will be added. You can edit the name of the group. You can use the arrow buttons from the Command bar in order to indent or outdent the report group. After creating the report group, the report is to be added under it.

You can assign the report group to the report. To do so, select the report from the **Reports** window and then choose the Browse button from the **Report Group** field; the **Select Report Group** dialog box will be displayed, as shown in Figure 8-15. In this dialog box, select the group under which you want to place the selected report and then choose the **Select** button; the report will be assigned to the selected group and the dialog box will be closed.

Creating Report Batch

Batch report group allows you to create the report batch. In a report batch, the reports are created as a combination of two or more report types. Report batch is very essential for a project as it provides the detailed information of a project. To create a report batch, select the **Batch Reports** option from the **Reports** flyout of

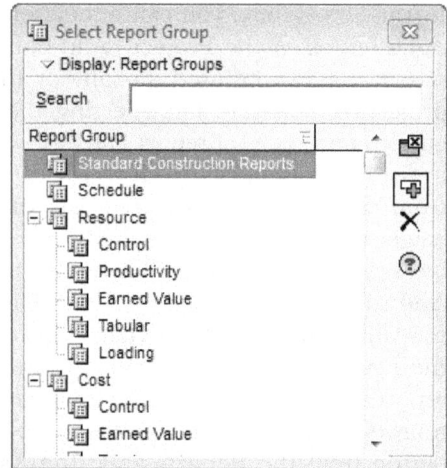

Figure 8-15 The Select Report Group dialog box

the **Tools** menu; the **Global Batch Reports** dialog box will be displayed, as shown in Figure 8-16. If you select the **Project** radio button from the **Global Batch Reports** dialog box, the name of dialog box will get modified to the **Project Batch Reports** dialog box with the reports of the project displayed. Note that by default the **Global Batch Reports** dialog box will be displayed on choosing the **Batch Reports** options from the **Reports** flyout of the **Tools** menu.

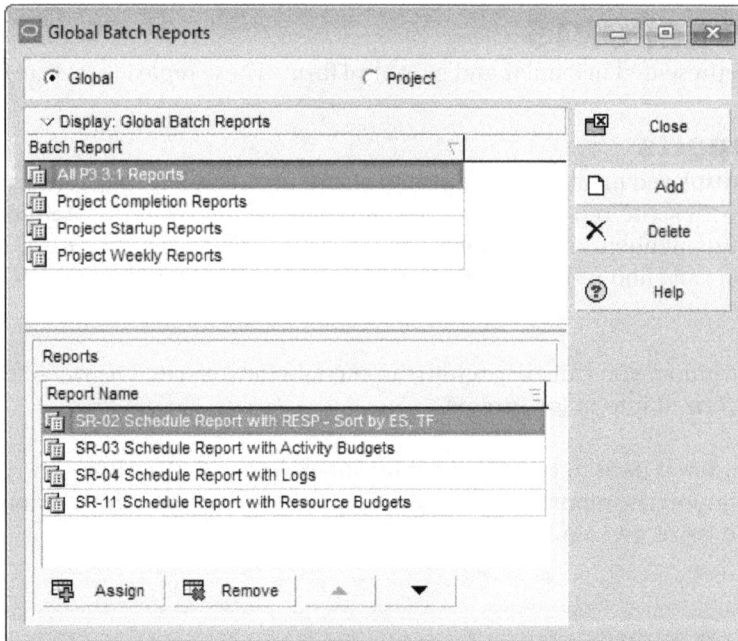

*Figure 8-16 The **Batch Reports** dialog box with the **Global** radio button selected*

In this dialog box, you can either select the **Global** radio button or the **Project** radio button to create a report that can be used globally or for a particular project, respectively. On selecting the **Global** radio button, four types of report batches will be displayed in the **Batch Report** area. If you select a batch report from the **Batch Report** area, the included report batches will be displayed in the **Reports** area. You can create your own types of batch reports by choosing the **Add** button from the Command bar. The newly created row will be added in the **Batch Report** area. Rename the newly created report batch and then assign the reports that will be included in that batch. To add the reports to the newly created batch, select the **Assign** button from the **Reports** area; the **Assign Reports** dialog box will be displayed, as shown in Figure 8-17. In this dialog box, select the reports that you want to assign for the batch and then choose the **Assign** button;

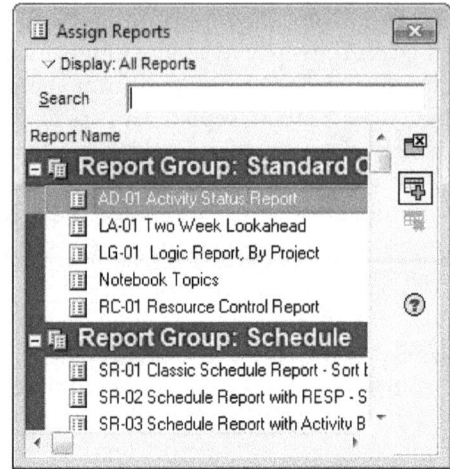

Figure 8-17 *The* ***Assign Reports*** *dialog box*

the report will be assigned for the newly created batch. Next, choose the **Close** button; the dialog box will be closed. In the **Reports** area, you can use the **Shift Up** or **Shift Down** buttons to shift the report up or down. Next, choose the **Close** button; the **Batch Reports** dialog box will be closed.

TYPES OF REPORTS

Reports can be represented in tabular and graphical form. These report forms are described next.

Tabular Reports

The reports are displayed in table format with the fields and detailed description that are needed in the report. To view the reports, choose the **Reports** option from the **Reports** flyout in the **Tools** menu of the menubar; the **Reports** window will be displayed, as shown in Figure 8-18. Alternatively, you can choose the **Reports** button from the Directory bar to invoke the **Reports** window.

In the **Reports** window, you can view reports in the tabular format. The default tabular report groups are categorized into many groups.

The **Reports** window is used to create, edit, run, and delete global and project reports. You can also export and import the reports to and from the Primavera P6. The description of the buttons in the Command bar is given in Table 8-1.

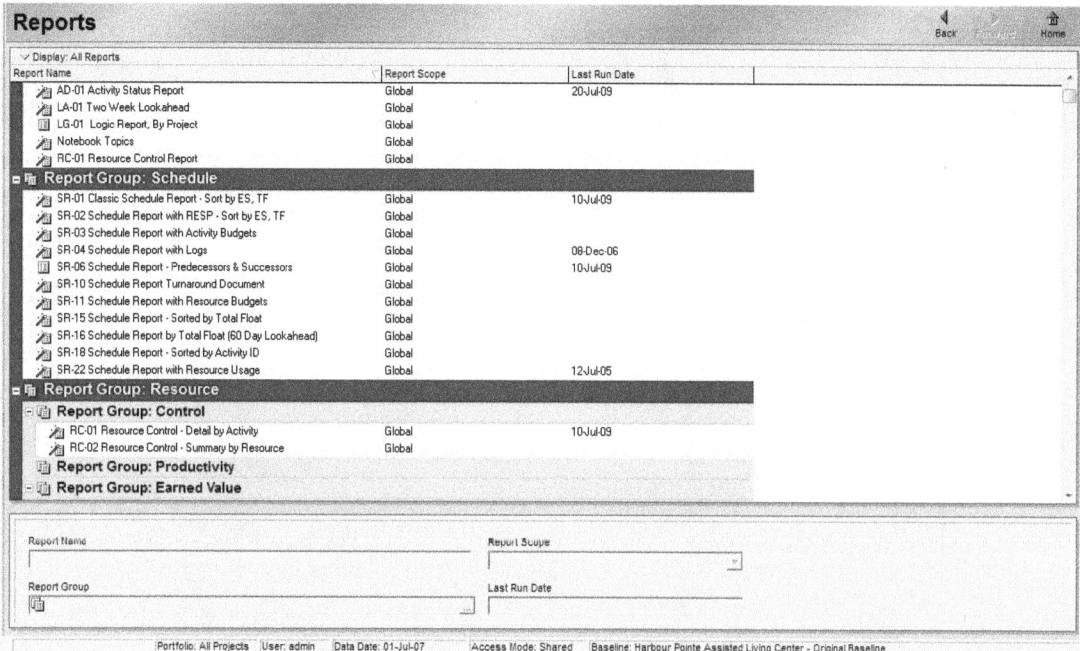

*Figure 8-18 The **Reports** window with the **Tabular** reports*

Table 8-1 The description of various buttons in the Command bar

Buttons	Description
Add	It allows you to create or add a report type to the selected report group.
Delete	It allows you to delete the report type from the selected report group.
Cut	It allows you to cut the report type from the desired report group and move it to the another report group.
Copy	It allows you to add the same report type from the desired report group to the another report group.
Paste	It allows you to paste the report type that you have copied or cut.
Modify	It allows you to modify the selected report in **Report Editor**.
Import	It allows you to import the report type from the Primavera P6.
Export	It allows you to export the report type from the selected group.
Wizard	It allows you to modify the wizard report.
Run Report	It allows you to view the report which is to be printed.
Run Batch	It allows you to run batch report for the selected project.

Run Report

To view the reports from the report group, you need to run the report from the group. To run the report, select the required report type under any report group head and then choose the **Run Report** button from the Command bar; the **Run Report** dialog box will be displayed, as shown in Figure 8-19.

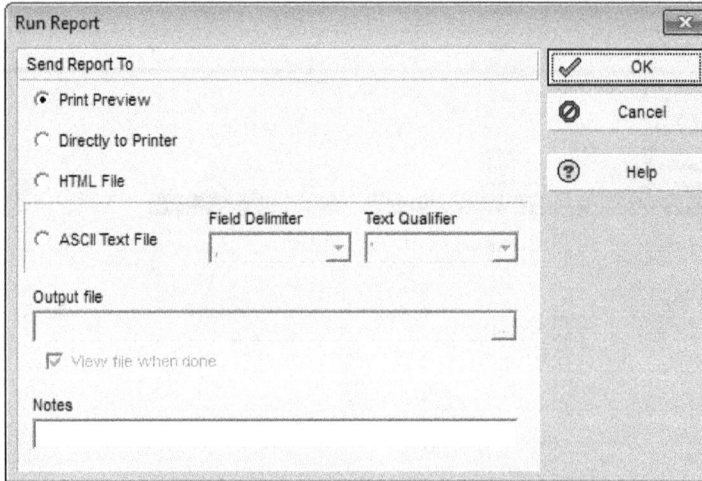

*Figure 8-19 The **Run Report** dialog box*

In this dialog box, you can define settings for the report to be displayed. In the **Send Report To** area of the dialog box, select the **Print Preview** radio button and then choose the **OK** button; the **Print Preview** window with the preview of the report will be displayed, refer to Figure 8-20.

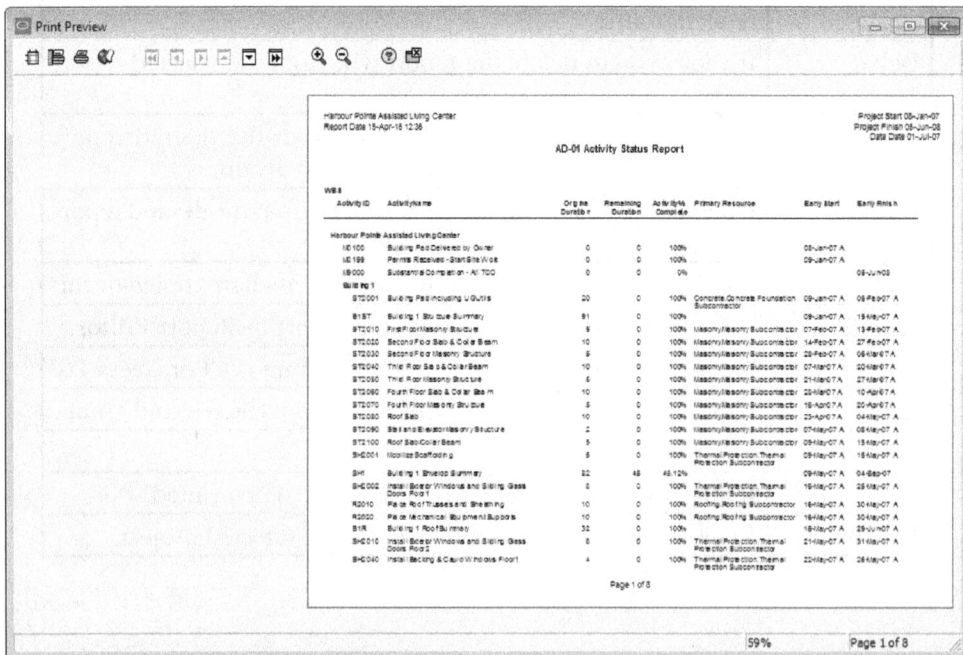

*Figure 8-20 The **Print Preview** window with the report preview*

In the **Run Report** dialog box, select the **Directly to Printer** radio button from the **Send Report To** area and then choose the **OK** button, the **Print** dialog box will be displayed, as shown in Figure 8-21.

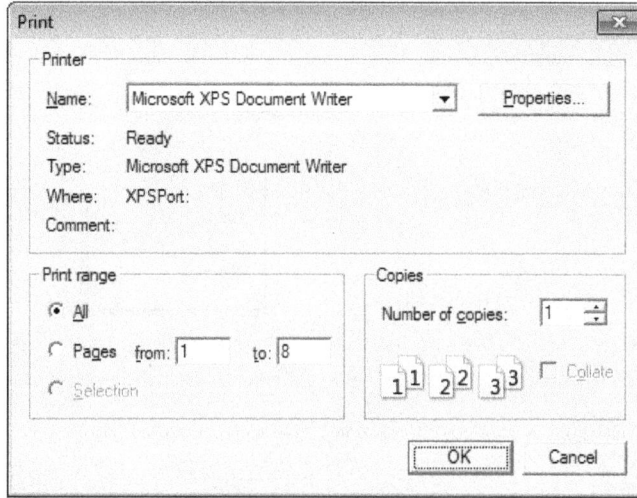

*Figure 8-21 The **Print** dialog box*

In the **Print** dialog box, you can select a printing device from the **Name** drop-down list of the **Printer** area. This dialog box displays the information regarding the selected printer in the **Status**, **Type**, **Where**, and **Comment** fields of the **Printer** area. The **Properties** button is used to set printer properties. The printer properties depend on the type of printer selected from the **Name** drop-down in the **Print** dialog box. If you choose the **Properties** button, a dialog box with the set of printer properties corresponding to the selected printer will be displayed.

Note
*If the PDF convertor is installed in your system then you can select the **Adobe PDF** option from the **Name** drop-down list in the **Print** dialog box to convert the report into PDF.*

The **Microsoft XPS Document Writer** option is selected in the **Name** drop-down lists. Now, on choosing the **Properties** button; the **Microsoft XPS Document Writer Document Properties** dialog box will be displayed with the **Layout** tab chosen. Figure 8-22 displays the options related to the orientation of the paper in the **Layout** tab. You can change the orientation of the paper from the **Orientation** area. Choose the **Advanced** button in this tab; the **Microsoft XPS Document Writer Advanced Options** dialog box will be displayed. In this dialog box, you can modify and set printing preferences, such as paper size, refer to Figure 8-23.

Note
*The appearance and options available in the **Print** dialog box depend upon the type of printer selected and the operating system used.*

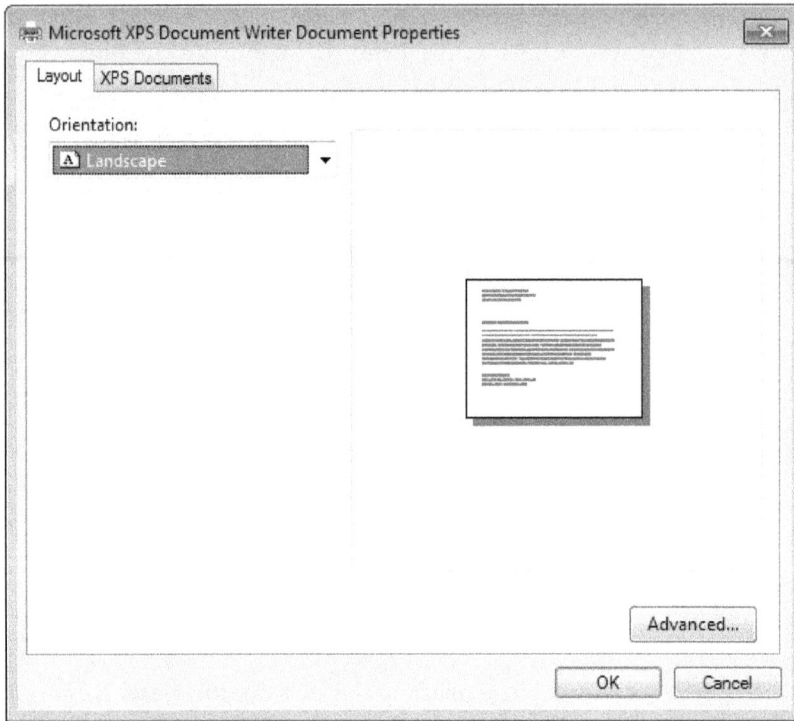

*Figure 8-22 The **Microsoft XPS Document Writer Document Properties** dialog box*

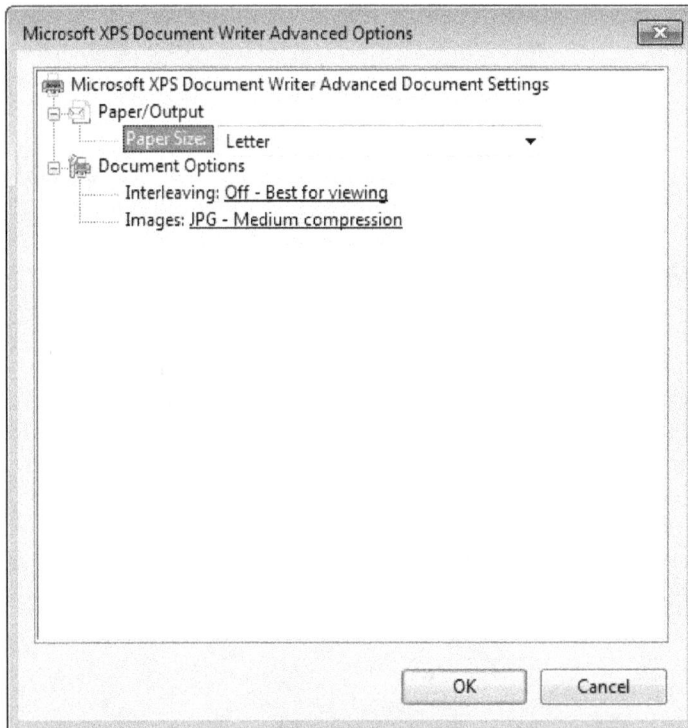

*Figure 8-23 The **Microsoft XPS Document Writer Advanced Options** dialog box*

The **Print Range** area in the **Print** dialog box has three radio buttons: **All**, **Pages**, and **Selection**. You can select any of these radio buttons to specify the parameters of the view to be printed.

Once the report is created, it can be previewed, printed or can be saved as a text or HTML file. Saving the report helps to import data to a spreadsheet, e-mail it, archive it or publish it on a website.

Run Batch

The batch run helps in running report in batches which implies that more than one report will be generated. As the name suggests the report is to be run in batches for example weekly, on completion or at start up phase. To run a batch report, choose the **Run Batch** button from the Command bar; the **Run Batch** dialog box will be displayed, as shown in Figure 8-24. In this dialog box, select the type of batch you want to run for generating the report. After selecting the desired report type from the dialog box, choose the **OK** button; the **Run Report** dialog box will be displayed with the report preview. Choose the **OK** button; the report generating process will start and the report in a batch of two or three will be generated.

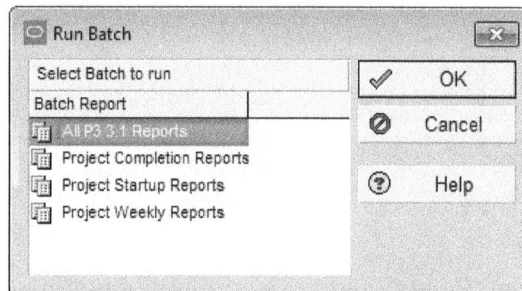

Figure 8-24 The **Run Batch** *dialog box*

Run Batch Report as a Job Service

The report can be generated in batches or as a single report. The reports generated in a batch can be according to the sequential arrangement of the service to be performed. To add the reports according to the job service, you need to choose the **Job Services** option from the **Tools** menu; the **Job Services** dialog box will be displayed, as shown in Figure 8-25.

In this dialog box, choose the **Add** button from the Command bar; a new row will be added in the **Job Queue** area. In the **Job #** column, specify the number according to which the job service is to be performed. You can type a brief description about the service under the **Job Name** column. To set the status of the report as either **Enabled** or **Disabled**, click under the **Status** column and then select the desired option. You can set the status as **Enabled** to activate the batch report service and you can suspend the job service anytime by selecting the **Disabled** option. In the **Service Type** column, you can specify the type of report that is added such a summarize, batch report, schedule type, and so on. The **Last Run** column will display the date on which the jobs were run previously. The **Next Run** column displays that date on which the jobs will be run in future.

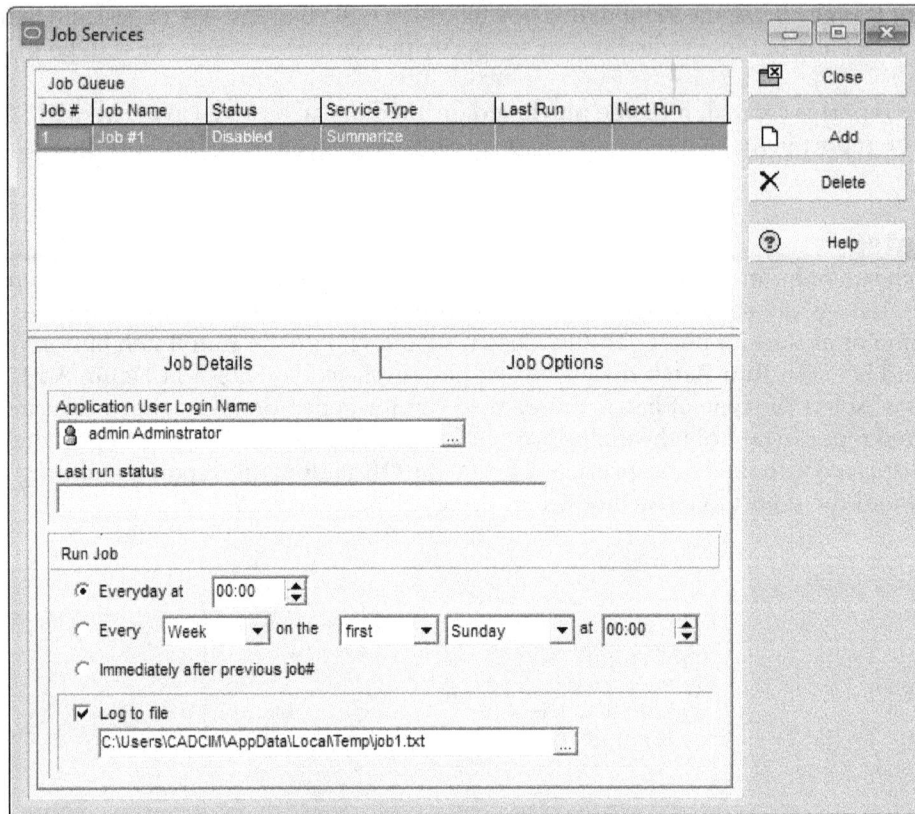

*Figure 8-25 The **Job Services** dialog box*

In the **Job Details** tab, you can specify the appropriate access rights to the job service. In order to do that, you need to choose the Browse button from the **Application User Login Name** field and assign the required user. The **Last run status** edit box displays the date on which the report was last run. In the **Run Job** area of the **Job Details** tab, arrange the schedule for the job service to be done such as everyday, every week or immediately after the previous job. You can select the required radio button according to which the job service will be done.

The job services that are to be executed on a project can be assigned in the **Job Options** tab of the **Job Services** dialog box. To assign job services to a project, you need to choose the **Assign** button from the **Job Options** tab; the **Add Projects** dialog box will be displayed. In this dialog box, select the project to which you want to assign the job services and choose the **Assign** button to assign the project to the job service and then choose the **Close** button to close the dialog box. Next, close the **Job Services** dialog box.

Graphical Reports
The project schedule can be represented graphically. There are two types of graphical reports: Histogram and S-curve.

Histogram Report

The Histogram report can be shown or calculated for the resources assigned to the activities. The report shown in histogram depicts the usage profile of the resources. These reports can be shown by using the **Resource Usage Profile** option from the **Activity** toolbar. When you choose this option, the **Resource Usage Profile** interface will be displayed, as shown in Figure 8-26. The **Activities** window is divided into four segments. The upper two segment shows the activities with the Gantt chart of a project. The lower left segment shows the resources and its usage profile. The settings of the usage profile can be done by using the **Resource Usage Profile Options** dialog box. To invoke this dialog box, right-click on the lower right segment of the **Activities** window; a shortcut menu will be displayed. Select the **Resource Usage Profile Options** option from the menu; the **Resources Usage Profile Options** dialog box will be displayed, as shown in Figure 8-27.

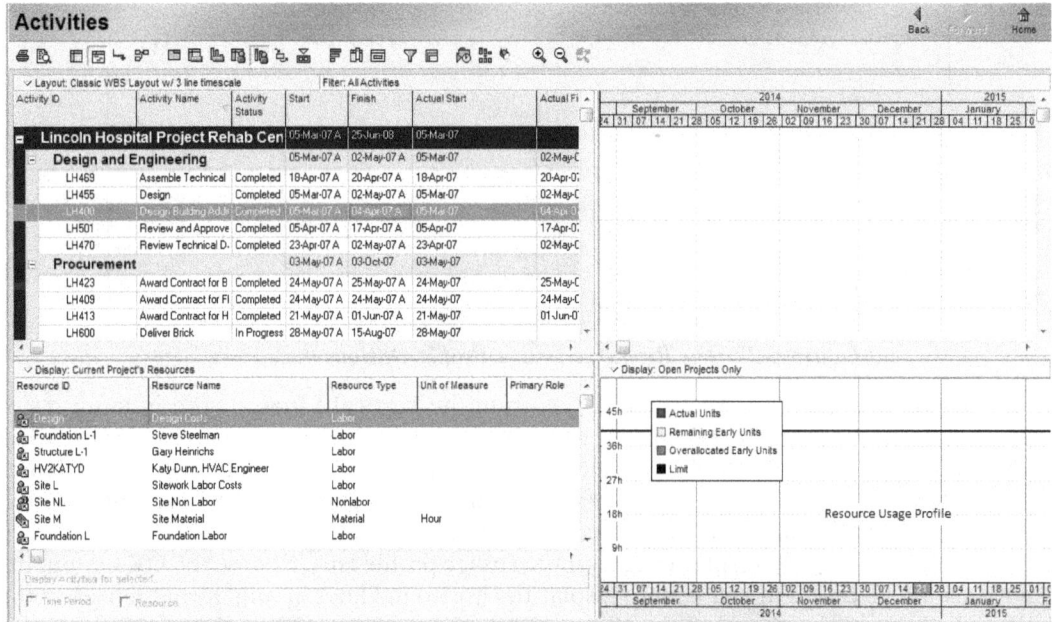

*Figure 8-26 The **Activities** window with the **Resource Usage Profile** interface*

In this dialog box, you can select the desired radio button from the **Display** area of the **Data** tab to display either the Units or the Cost data in the usage profile. In the **Show Bars/Curves** area, select the check box that you want to display for the resources. You can select the check boxes corresponding to the **Budgeted**, **Actual**, **Remaining Early** and **Remaining Late** options under **By Date** or **Cumulative** columns according to which you want to show these bars and curves. You can change the color of bars by choosing the button from the **Color** column. The remaining resource can be displayed by choosing the **Total Remaining** radio button. The early remaining will be shown in green solid color whereas the late remaining will be shown in a pattern in green color. You can change these colors by choosing the color buttons corresponding to the **Budgeted** and **Actual** options under the **Color** column. You can select the **Show Limit** check box to display the limit of the resources in the usage profile. By selecting the **Show Overallocation** check box, you can show the overallocation of resources if any.

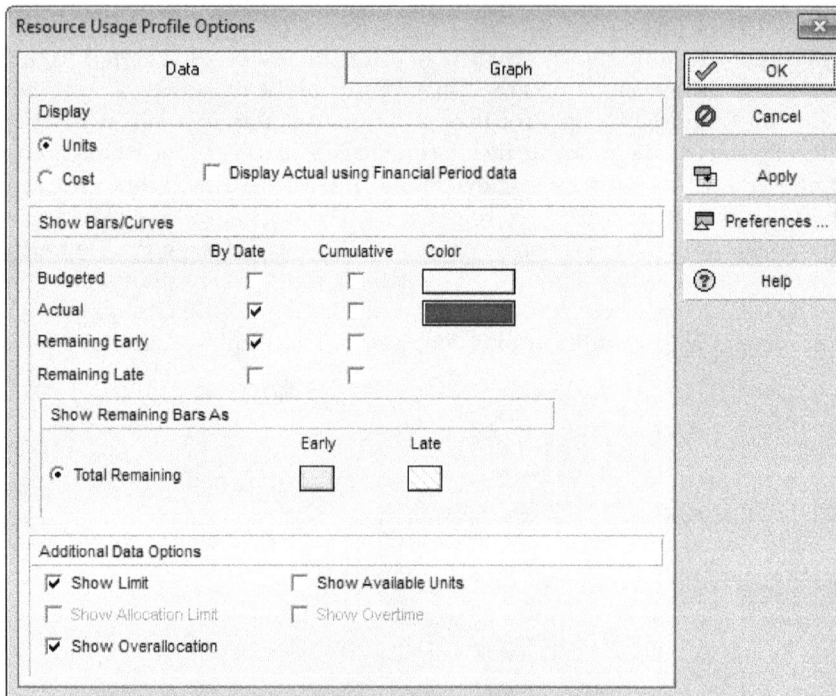

*Figure 8-27 The **Resource Usage Profile Options** dialog box*

In the **Graph** tab, select the **Major** check box from the **Vertical Lines** area to indicate major time units in the form of vertical lines in graph. Similarly, for minor time units you can select the **Minor** check box. Select the **Dotted** check box to display the horizontal lines in dotted form in grey color. You can change the color of horizontal lines by choosing the **Line Color** button from the **Horizontal Lines** area. On doing so, the **Color** dialog box will be displayed. You can select the required color to display in the resource usage profile area. Choose the **OK** button; the dialog box will be closed. If you want to display the horizontal lines in solid form then select the **Solid** radio button. You can select the **None** radio button to remove those lines from the area.

Select the **Show Legend** check box from the **Additional Display Options** area to display the legend with color in the current layout. If you want to display graph bars in three dimensions then select the **3D Bars** check box. You can set the color for the background by using the **Background** color button. You can select the **Calculate Average** check box to specify the values you want to use to divide the timescale intervals. As you select the **Calculate Average** check box, the other options below it will be activated. You can specify the value in the **Divide interval totals by** edit box by which you want the timescale interval to be divided. If you select the **Base on Hours per Time Period** check box, division increments specified are displayed in the **User Preferences** dialog box for the corresponding date interval. For example 1h corresponds to Hour day interval, 2h corresponds to Shift day interval, and 8h corresponds to the day date interval. In the **Unit of Measure** edit box, you can enter the unit to measure the timescale interval. Choose the **OK** button from the Command bar; the dialog box will be closed.

Note

*The settings for the time period can be done by choosing the **Preferences** button from the Command bar in the **Resources Usage profile Options** dialog box.*

S-Curve

The other type of graphical representation of a report is in the form of S-curve. This curve is in S shape and shows the pattern in which the money is spent for the resources. To generate the S-Curve, right-click in the **Resources Usage Profile** option; the **Resource Usage Profile Options** dialog box will be displayed, refer to Figure 8-27. In this dialog box, you can set the required setting for the curve to be displayed.

PRINTING LAYOUTS

After creating a report, you need to print the project data. Printing settings include page orientation, margins, and header/footer setting. You can also view the preview for the layout and reports.

Page Settings

You can customize the printed layouts and the reports to be represented as data in number of ways. You can set the header and footer settings, margins, orientation, scaling, and paper size. To setup the page for printing the report, select the **Page Setup** option from the **File** menu; the **Page Setup** dialog box with the **Page** tab chosen will be displayed, as shown in Figure 8-28.

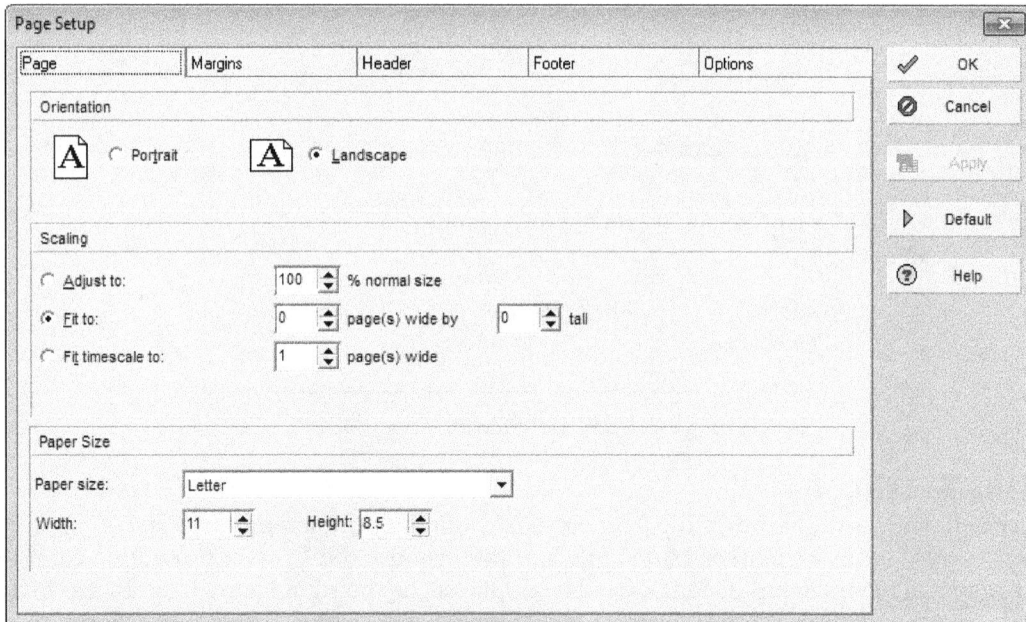

*Figure 8-28 The **Page Setup** dialog box*

The tabs contained in the **Page Setup** dialog box will allow you to specify the settings for the page. You can set the orientation of the page either in portrait or landscape by choosing the required radio button in the **Orientation** area. To set the scaling of the page, you can select the

Adjust to radio button and set the spinner to certain required percentage. To set the page by defining the width and height, select the **Fit to** radio button. To set the paper size, select the required option from the **Paper size** drop-down list. The **Width** and **Height** spinners will show the width and height of the paper depending upon the option selected from the **Paper size** drop-down list.

The options in the **Margins** tab allow you to set the top, bottom, left, and right margins by entering the required values corresponding to the spinners.

The options in the **Header** tab are used to create a custom header for the layout or report. The options in the **Divided into: Section** rollout are used to enter the value by which the header will be divided. On specifying the value in the **Divided into: Sections** rollout, the section in the **Define header** area will get divided accordingly. You can select the **Show Section Divider Lines** check box to show the divider lines in the report. You can select the required option from the **Include on** drop-down list for which you want these settings to include. In the **Add Text** area, you can write a text that can be included in the header. Rest of the formatting can be done using the tools given in the **Add Text** area. You can add a logo to the header. To do so, choose the **Picture** button from the **Add Text** area; the **Picture** dialog box will be displayed, as shown in Figure 8-29. In this dialog box, you can choose the Browse button in the **Picture Source** edit box; the **Picture** dialog box will be displayed and will allow you to browse to the path of the saved picture. Choose the **Open** button; the path will be displayed in the **Picture Source** edit box. You can type the alternate text for the logo in the **Alternate Text** area. Next, choose the **OK** button; the **Picture** dialog box will be closed.

The options in the **Footer** tab allow you to specify settings for the footer as done earlier in the **Header** tab. The options in the **Options** tab allow you to specify page settings and further print setting.

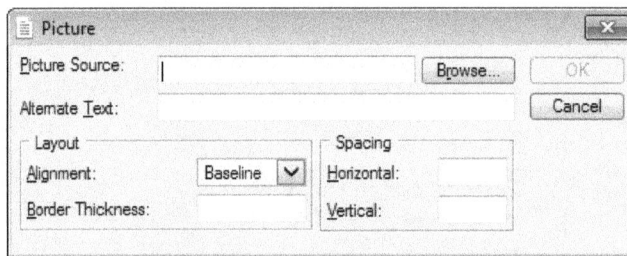

*Figure 8-29 The **Picture** dialog box*

Print Settings

You can define settings for printing a report or a layout. To print a report or a layout, open the required view in the Primavera P6 module and then choose the **Print Setup** option from the **File** menu; the **Print Setup** dialog box will be displayed, as shown in Figure 8-30. In this dialog box, select the name of the printer from the **Name** drop-down list. The paper size can be set earlier while doing the page set up or can be done in the **Paper** area of the **Print Setup** dialog box. The orientation of the print can be selected from the **Orientation** area.

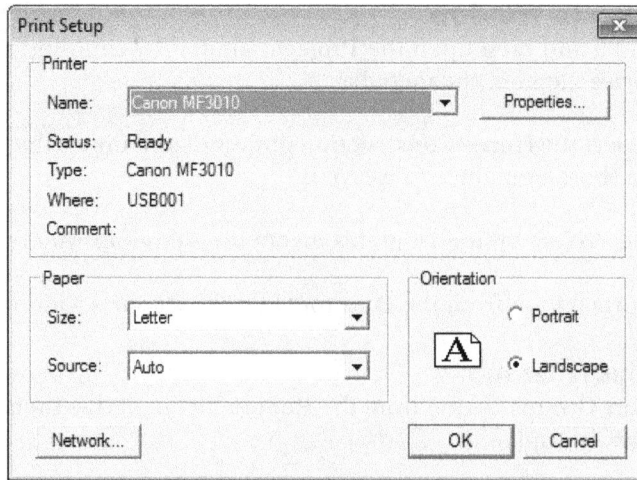

Figure 8-30 The **Print Setup** *dialog box*

TUTORIALS

To perform the tutorials of this chapter, you need to complete Tutorial 1 and Tutorial 2 of Chapter 2.

To perform this tutorial you need *c07_Primavera_P6_tut.zip* file. You can download this file from *www.cadcim.com* using these steps:

1. To download the file browse to *Textbooks > Civil/GIS> Primavera P6 > Exploring Oracle Primavera P6 v7.0*. Next, select the *c07_Primavera_P6_tut.zip* file from the **Tutorial Files** drop-down list. Choose the corresponding **Download** button to download the data file.

2. Now, save and extract the downloaded folder to the following location:

    C:/PM6

3. Import the *c07_CONS_Home_tut01* file from the extracted folder to the Construction EPS created in Tutorial 1 of Chapter 2.

Tutorial 1   Report Group and Run Report

In this tutorial, you will create a report group for the existing **Home Construction** project. You will also create the report and run that report for the project.   **(Expected time: 50 min)**

The following steps are required to complete this tutorial:

a. Open the **Reports** window.
b. Create report group using the **Report Group** option.
c. Create and run report.
d. Export the project.

Opening the Reports Window

1. Open Primavera P6 and then open the **Projects** window by choosing the **Projects** option from the **Enterprise** menu in the menubar.

2. In this window, select the **Home Construction** project placed under the **Construction** node and right-click; a shortcut menu is displayed.

3. Choose the **Open Project** option from this menu; the **Activities** window is displayed.

4. Choose the **Reports** button from the Directory bar; the **Reports** window is displayed.

Creating the Report Group

1. Choose the **Report Groups** option from the **Reports** flyout of the **Tools** menu; the **Report Groups** dialog box is displayed.

2. Choose the **Add** button; a row is added with the name **New Report Group**.

3. Rename the project as **Home Construction Project** and then press ENTER, refer to Figure 8-31.

4. Choose the **Shift Up** button to shift this report group on the top, as shown in Figure 8-31.

5. Choose the **Close** button; the dialog box is closed and the **Home Construction** project is displayed in the **Reports** window.

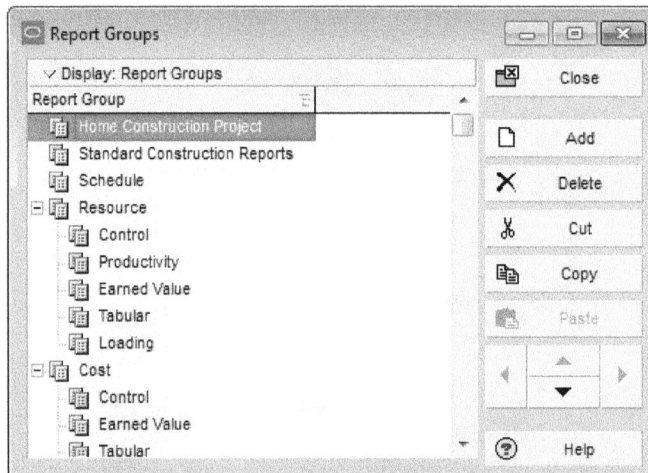

*Figure 8-31 The **Report Groups** dialog box*

Creating and Running the Report

1. Select the **Report Group: Home Construction Project** head in the **Reports** window if it is not selected.

2. Choose the **Add** button; **Report Wizard** with the **Create or Modify Report** page and with the **New Report** radio button selected.

3. Choose the **Next** button; the **Select Subject Area** page is displayed. In this page, select the **Activities** option.

4. Choose the **Next** button; the **Select Additional Subject Areas** page is displayed. Ensure that the **Activities** option is selected. Choose the **Next** button; the **Configure Selected Subject Areas** page is displayed.

5. Choose the **Columns** button; the **Columns** dialog box is displayed.

6. In this dialog box, expand the **General** node and shift the **Activity Name** option from the **Available Options** area to the **Selected Options** area using the **Add to list** button.

7. Expand the **Dates** node and then add the **BL1 Start, BL1 Finish, Actual Start**, and **Actual Finish** options to the **Selected Options** area.

8. Expand the **Percent Completes** node and then shift the **Activity % Complete** option to the **Selected Options** area. Also, arrange the options, as shown in Figure 8-32.

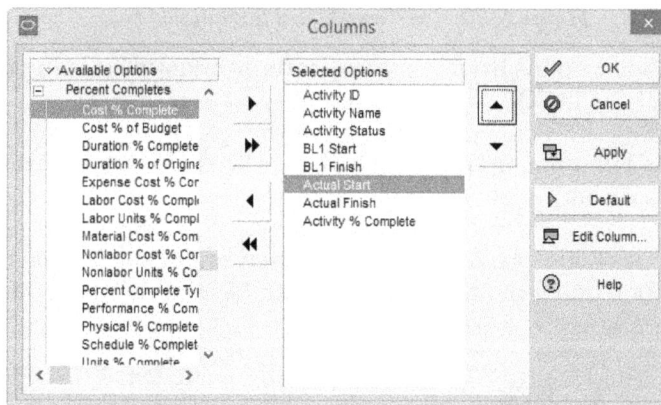

*Figure 8-32 The **Columns** dialog box with the arranged options*

9. Choose the **OK** button; the **Columns** dialog box is closed and the columns for reports are customized.

10. Choose the **Group & Sort** button; the **Group and Sort** dialog box is displayed.

11. In this dialog box, click under the **Group By** column and select the **WBS** option from the list displayed. On selecting this option, other fields are filled automatically, as shown in Figure 8-33.

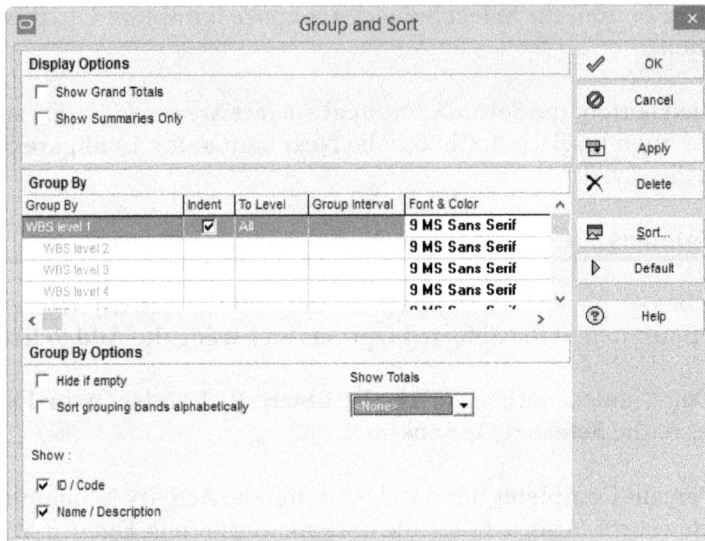

Figure 8-33 The **Group and Sort** *dialog box*

12. Choose the **<None>** option from the **Show Totals** drop-down list in the **Group By Options** area and then choose the **OK** button; the **Group and Sort** dialog box is closed.

13. Now, choose the **Filter** button from the **Configure Selected Subject Areas** page; the **Filter** dialog box is displayed.

14. In this dialog box, select the **Activity Type** option under the **Parameter** column, the **is not equal to** option under the **Is** column, and the **Level of Effort** option under the **Value** column, as shown in Figure 8-34.

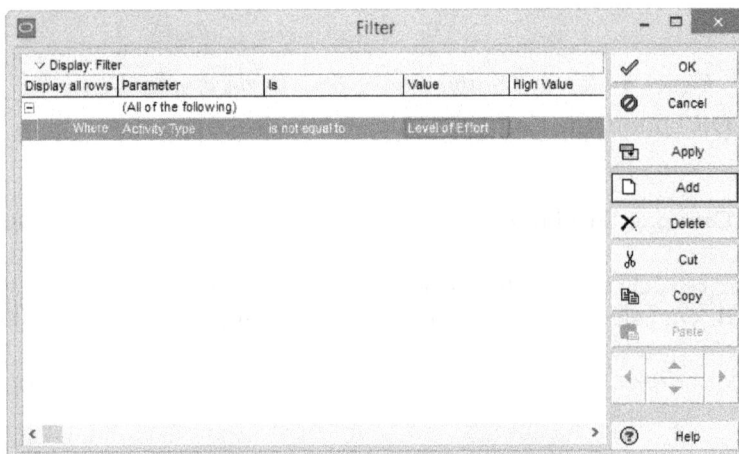

Figure 8-34 The **Filter** *dialog box*

15. Next, choose the **OK** button; the **Filter** dialog box is closed.

16. Choose the **Next** button from the **Configure Selected Subject Areas** page; the **Report Title** page is displayed. In this page, enter the title as **Activity Status and Progress Report** in the edit box.

17. Choose the **Next** button; the **Report Generated** page is displayed. Select the **Run Report** button from this page; the **Run Report** dialog box is displayed, as shown in Figure 8-35.

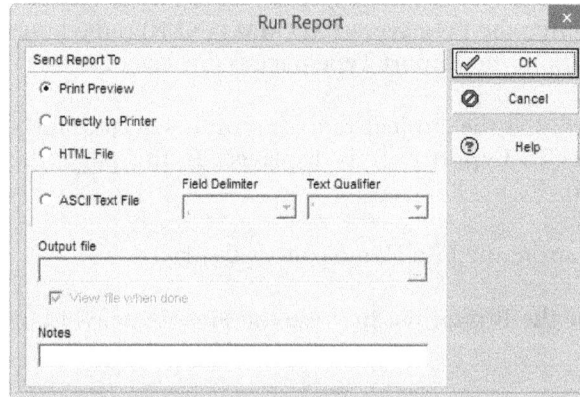

*Figure 8-35 The **Run Report** dialog box*

18. Ensure that the **Print Preview** radio button is selected and then choose the **OK** button; the **Print Preview** window with the preview of report is displayed. Close the **Print Preview** window and then choose the **Next** button from the **Report Generated** page; the **Congratulations** page is displayed.

19. In this page, choose the **Save Report** button; the **Primavera P6** message box is displayed.

20. Choose the **OK** button and then choose the **Finish** button; **Report Wizard** is closed and the report is added under the **Home Construction Project** report group.

Printing the Report

1. Select the report and then choose the **Run Report** button from the Command bar; the **Run Report** dialog box is displayed, refer to Figure 8-35.

2. In this dialog box, select the **ASCII Text File** radio button and then select ',' (comma) from the **Field Delimiter** drop-down list and ' (apostrophe) from the **Text Qualifier** drop-down list.

3. Choose the Browse button from the **Output file** field; the **Save report as** dialog box is displayed.

4. Browse to *C:\PM6* and create a folder with the name **c08** and then enter **Activity Status and Project Report** in the **File name** edit box and choose the **Save** button; the dialog box is closed and the report is saved.

5. Ensure that **.csv** is entered at the end of the path given in the **Output file** field and then choose the **OK** button; the **Run Report** dialog box is closed.

Exporting the Project

You need to export projects to save the project file for future reference. Ensure that the project to be exported is opened in Primavera P6.

1. To export a project, choose the **Export** option from the **File** menu; the **Export** wizard with the **Export Format** page is displayed.

2. In this page, ensure that the **Primavera PM / MM - (XER)** radio button is selected and then choose the **Next** button; the **Export Type** page is displayed.

3. In this page, ensure that the **Project** radio button is selected and then choose the **Next** button; the **Projects To Export** page is displayed. In this page, the **Home Construction** project is displayed.

4. Choose the **Next** button; the **File Name** page is displayed.

5. In this page, choose the Browse button from the **File Name** edit box; the **Save File** dialog box is displayed.

6. In this dialog box, browse to the *C:\PM6* and create a folder with a name **c08**. Then, save the file as **c08_CONS_HOME_tut01** and then choose the **Save** button.

7. Choose the **Finish** button; the **Primavera P6** message box is displayed with the message that the export was successful.

8. Choose the **OK** button; the file gets exported.

Tutorial 2   Resource Report

In this tutorial, you will run an existing report for the **Home Construction** project.

(Expected time: 40 min)

The following steps are required to complete this tutorial:

a. Open the **Reports** window.
b. Create and run the report.
c. Print report.
d. Export the project.

Opening the Reports Window

1. Open Primavera P6 and then open the **Projects** window by choosing the **Projects** option from the **Enterprise** menu in the menubar.

2. In this window, select the **Home Construction** project placed under the **Construction** EPS node and right-click; a shortcut menu is displayed.

3. Choose the **Open Project** option from this menu; the **Activities** window is displayed.

4. Choose the **Reports** button from the Directory bar; the **Reports** window is displayed.

Creating and Running the Report

1. Select the **Report Group: Home Construction Project** if not selected in the **Reports** window.

2. Choose the **Add** button from the Command bar; **Report Wizard** with the **Create or Modify Report** page and with the **New Report** radio button selected is displayed.

3. Choose the **Next** button; the **Select Subject Area** page is displayed.

4. In this page, select the **Resources** option and then choose the **Next** button; the **Select Additional Subject Areas** page is displayed. Ensure that the **Resources** option is selected in this page.

5. Choose the **Next** button; the **Configure Selected Subject Areas** page is displayed. Choose the **Group & Sort** button; the **Group and Sort** dialog box is displayed.

6. In this dialog box, select the **Show Grand Totals** check box from the **Display Options** area and then choose the **OK** button.

7. Choose the **Filter** button from the **Configure Selected Subject Areas** page; the **Filter** dialog box is displayed.

8. In this dialog box, select the **Assigned to Current Project** option under the **Parameter** column and ensure that the **equals** and **Yes** options are selected under the **Is** and **Value** columns, respectively.

9. Choose the **OK** button; the **Filter** dialog box is closed.

10. Choose the **Next** button; the **Report Title** page is displayed. In this page, enter the title as **Resource Report**.

11. Choose the **Next** button twice; the **Congratulations** page is displayed.

12. In this page, choose the **Save Report** button; the **Primavera P6** dialog box is displayed.

13. Choose the **OK** button and then choose the **Finish** button; **Report Wizard** is closed and a report is added under the **Home Construction Project** report group.

Printing the Report

1. Select the report and then choose the **Run Report** button from the Command bar; the **Run Report** dialog box is displayed.

2. In this dialog box, select the **ASCII Text File** radio button and then select **,** (comma) from the **Field Delimiter** drop-down list and **'** (apostrophe) from the **Text Qualifier** drop-down list.

3. Choose the Browse button from the **Output file** field; the **Save report as** dialog box is displayed.

4. Browse to C:\PM6\c08 and then enter the report name as **Resource Report** in the **File name** edit box and then choose the **Save** button; the dialog box is closed and the report is saved.

5. Ensure that the **.csv** is entered at the end of the path given in the **Output file** field and then choose the **OK** button; the **Run Report** dialog box is closed.

Exporting the Project

You need to export projects to save the project file for future reference. Ensure that the project to be exported is opened in Primavera P6.

1. To export the project, choose the **Export** option from the **File** menu; the **Export** wizard with the **Export Format** page is displayed.

2. In this page, ensure that the **Primavera PM/MM - (XER)** radio button is selected and then choose the **Next** button; the **Export Type** page is displayed.

3. In this page, ensure that the **Project** radio button is selected and then choose the **Next** button; the **Projects To Export** page is displayed. In this page, the **Home Construction** project is displayed.

4. Choose the **Next** button; the **File Name** page is displayed.

5. In this page, choose the Browse button from the **File Name** edit box; the **Save File** dialog box is displayed.

6. In this dialog box, browse to the *C:\PM6\c08* and save the file with the name **c08_CONS_HOME_tut02** and then choose the **Save** button.

7. Choose the **Finish** button; the **Primavera P6** message box is displayed with the message that the export was successful.

8. Choose the **OK** button; the file is exported.

Self-Evaluation Test

Answer the following questions and then compare them to those given at the end of this chapter:

1. Which of the following pages of **Report Wizard** is used to select the subject area?

      (a) **Select Additional Subject Area**   (b) **Configure Selected Subject Area**
      (c) **Select Subject Area**   (d) **Date Options**

2. Which of the following pages is used to configure column, group, and filter of report?

    (a) **Create or Modify Report**   (b) **Report title**
    (c) **Configure Selected Subject Area**   (d) **Select Subject Area**

3. Which of the following pages is used to define setting for the time related report?

    (a) **Date Options**   (b) **Select Subject Area**
    (c) **Report Generated**   (d) **Congratulations**

4. Which of the following options is used to invoke the **Run Report** dialog box?

    (a) **Run Report**   (b) **Run Batch**
    (c) **Modify Report**   (d) **Modify Wizard**

5. The reports can be created or modified in the _____ wizard.

6. The reports can be filtered using the _____ dialog box.

7. The _____ page is used to define time settings for the report.

8. There are two types of reports that can be generated and displayed for the project. (T/F)

9. A batch report allows you to create reports in a group of two only. (T/F)

10. You can modify reports using the **Modify** tool. (T/F)

Review Questions

Answer the following questions:

1. The report groups can be created by choosing the **Report Group** option from the _____ menu.

2. The _____ window is used to create, edit, run, and delete reports.

3. The _____ button is used to export the report type from a group.

4. To show the over allocation of resources, the _____ check box is to be selected.

5. The _____ tab is used for define page settings.

6. **Report Editor** allows you to edit a report. (T/F)

7. The **Group & Sort** dialog box helps to filter the report. (T/F)

8. You can run the report in the job sequence order by adding the reports in the **Run Batch** dialog box. (T/F)

9. You can choose the **Run Report** button from the **Run Generated** page. (T/F)

10. You can select the **Time Distributed Data** check box from the **Select Subject Area** page. (T/F)

EXERCISE

Exercise 1

In this exercise, you will create a report group for the **Hospital Building** project and also create a report for the Activity Project status.   **(Expected time: 50 min)**

Index

Other Publications by CADCIM Technologies

The following is the list of some of the publications by CADCIM Technologies. Please visit *www.cadcim.com* for the complete listing.

Raster Design Textbook
- Exploring Raster Design 2016

Autodesk Revit Architecture Textbooks
- Autodesk Revit Architecture 2016 for Architects and Designers, 12th Edition
- Autodesk Revit Architecture 2015 for Architects and Designers, 11th Edition

AutoCAD Civil 3D Textbooks
- Exploring AutoCAD Civil 3D 2016, 6th Edition
- Exploring AutoCAD Civil 3D 2015, 5th Edition

AutoCAD Map 3D Textbooks
- Exploring AutoCAD Map 3D 2016, 6th Edition
- Exploring AutoCAD Map 3D 2015, 5th Edition

Autodesk Revit Structure Textbooks
- Exploring Autodesk Revit Structure 2016, 6th Edition
- Exploring Autodesk Revit Structure 2015, 5th Edition

Autodesk Revit MEP Textbooks
- Exploring Autodesk Revit MEP 2016, 3rd Edition
- Exploring Autodesk Revit MEP 2015

Autodesk Navisworks Textbooks
- Exploring Autodesk Navisworks 2016, 3rd Edition
- Exploring Autodesk Navisworks 2015

AutoCAD MEP Textbooks
- AutoCAD MEP 2016 for Designers
- AutoCAD MEP 2015 for Designers

SOLIDWORKS Textbooks
- SOLIDWORKS 2016 for Designers, 14th Edition
- SOLIDWORKS 2015 for Designers, 13th Edition

EdgeCAM Textbooks
- EdgeCAM 11.0 for Manufacturers
- EdgeCAM 10.0 for Manufacturers

CATIA Textbooks
• CATIA V5-6R2015 for Designers, 13th Edition
• CATIA V5-6R2014 for Designers, 13th Edition

Creo Parametric and Pro/ENGINEER Textbooks
• PTC Creo Parametric 3.0 for Designers, 3rd Edition
• Creo Parametric 2.0 for Designers

Autodesk Alias Textbooks
• Learning Autodesk Alias Design 2016, 5th Edition
• Learning Autodesk Alias Design 2015

Autodesk Simulation Mechanical Textbook
• Autodesk Simulation Mechanical 2014

ANSYS Textbooks
• ANSYS Workbench 14.0: A Tutorial Approach
• ANSYS 11.0 for Designers

Customizing AutoCAD Textbook
• Customizing AutoCAD 2013

AutoCAD LT Textbooks
• AutoCAD LT 2016 for Designers, 11th Edition
• AutoCAD LT 2015 for Designers, 10th Edition

AutoCAD Electrical Textbooks
• AutoCAD Electrical 2016 for Electrical Control Designers, 7th Edition
• AutoCAD Electrical 2015 for Electrical Control Designers, 6th Edition

Autodesk Inventor Textbooks
• Autodesk Inventor 2016 for Designers, 16th Edition
• Autodesk Inventor 2015 for Designers, 15th Edition

Solid Edge Textbooks
• Solid Edge ST8 for Designers, 13th Edition
• Solid Edge ST7 for Designers, 12th Edition

NX Textbooks
• NX 10.0 for Designers, 9th Edition
• NX 9.0 for Designers, 8th Edition

3ds Max Design Textbooks
• Autodesk 3ds Max Design 2015: A Tutorial Approach, 15th Edition
• Autodesk 3ds Max Design 2014: A Tutorial Approach

3ds Max Textbooks
- Autodesk 3ds Max 2016: A Comprehensive Guide, 16th Edition
- Autodesk 3ds Max 2016 for Beginners: A Tutorial Approach, 16th Edition
- Autodesk 3ds Max 2015: A Comprehensive Guide, 15th Edition

Maya Textbooks
- Autodesk Maya 2016: A Comprehensive Guide, 8th Edition
- Autodesk Maya 2015: A Comprehensive Guide, 7th Edition

Fusion Textbooks
- Blackmagic Design Fusion 7.0 Studio: A Tutorial Approach
- The eyeon Fusion 6.3: A Tutorial Approach

Computer Programming Textbooks
- Learning Oracle 11g: A PL/SQL Approach
- Learning ASP.NET AJAX

Paper Craft Book
- Constructing 3-Dimensional Models: A Paper-Craft Workbook

AutoCAD Textbooks Authored by Prof. Sham Tickoo and Published by Autodesk Press
- AutoCAD: A Problem-Solving Approach: 2013 and Beyond
- AutoCAD 2012: A Problem-Solving Approach
- AutoCAD 2011: A Problem-Solving Approach

Textbooks Authored by CADCIM Technologies and Published by Other Publishers

3D Studio MAX and VIZ Textbooks
- Learning 3ds max5: A Tutorial Approach
  (Complete manuscript available for free download on www.cadcim.com)
- Learning 3ds Max: A Tutorial Approach, Release 4
  Goodheart-Wilcox Publishers (USA)

CADCIM Technologies Textbooks Translated in Other Languages

3ds Max Textbook
- 3ds Max 2008: A Comprehensive Guide (Serbian Edition)
  Mikro Knjiga Publishing Company, Serbia

SolidWorks Textbooks
- SolidWorks 2008 for Designers (Serbian Edition)
  Mikro Knjiga Publishing Company, Serbia

- SolidWorks 2006 for Designers (Russian Edition)
  Piter Publishing Press, Russia
- SolidWorks 2006 for Designers (Serbian Edition)
  Mikro Knjiga Publishing Company, Serbia
- SolidWorks 2006 for Designers (Japanese Edition)
  Mikio Obi, Japan

NX Textbooks
- NX 6 for Designers (Korean Edition)
  Onsolutions, South Korea
- NX 5 for Designers (Korean Edition)
  Onsolutions, South Korea

Pro/ENGINEER Textbooks
- Pro/ENGINEER Wildfire 4.0 for Designers (Korean Edition)
  HongReung Science Publishing Company, South Korea
- Pro/ENGINEER Wildfire 3.0 for Designers (Korean Edition)
  HongReung Science Publishing Company, South Korea

AutoCAD Textbooks
- AutoCAD 2006 (Russian Edition)
  Piter Publishing Press, Russia
- AutoCAD 2005 (Russian Edition)
  Piter Publishing Press, Russia

Coming Soon from CADCIM Technologies
- Exploring Primavera P6 v8.4
- Exploring Autodesk Revit Architecture 2017 for Architects and Designers, 13th Edition
- Exploring Autodesk Revit Structure 2017, 7th Edition
- Exploring Autodesk Revit MEP 2017, 4th Edition
- NX Nastran 9.0 for Designers

Online Training Program Offered by CADCIM Technologies
CADCIM Technologies provides effective and affordable virtual online training on animation, architecture, and GIS softwares, computer programming languages, and Computer Aided Design and Manufacturing (CAD/CAM) software packages. The training will be delivered `live' via Internet at any time, any place, and at any pace to individuals, students of colleges, universities, and CAD/CAM training centers. For more information, please visit the following link: http://*www.cadcim.com.*

www.ingramcontent.com/pod-product-compliance
Lightning Source LLC
Chambersburg PA
CBHW061356210326
41598CB00035B/5999